RISING STARS
Mathematics

Year
6

Practice Book

Author: Paul Broadbent

ISBN: 978-1-78339-819-5
Text, design and layout © Rising Stars UK Ltd 2016

First published in 2016 by
Rising Stars UK Ltd, part of Hodder Education,
An Hachette UK Company
Carmelite House
50 Victoria Embankment
London EC4Y 0DZ
www.risingstars-uk.com

Author: Paul Broadbent
Programme consultants: Caroline Clissold, Cherri Moseley, Paul Broadbent
Publishers: Fiona Lazenby and Alexandra Riley
Editorial: Aidan Gill, Denise Moulton
Answer checker: Deborah Dobson
Project manager: Sue Walton

Series and character design: Steve Evans
Text design: Words & Pictures Ltd
Illustrations by Steve Evans

Cover design: Steve Evans and Words & Pictures Ltd

Printed by Liberduplex, Barcelona
A catalogue record for this title is available from the British Library.

FSC
www.fsc.org
MIX
Paper from responsible sources
FSC® C109440

"Hi there, it's me, Ali! Tick off each unit in the checklist below as you complete the practice questions. Then you will be able to see how much you have learnt this year!"

Unit	Title	I have completed…	
1	Whole and part numbers	1a	
		1b	
		1c	
		1d	
2	Calculations and algebra	2a	
		2b	
		2c	
3	Larger numbers	3a	
		3b	
		3c	
		3d	
4	2-D shapes, 3-D shapes and nets	4a	
		4b	
		4c	
5	Numbers in everyday life	5a	
		5b	
6	Solving problems	6a	
		6b	
		6c	
		6d	
7	Let's explore fractions and algebra!	7a	
		7b	
		7c	
		7d	
8	Using what you know	8a	
		8b	
		8c	
		8d	
9	Shapes and coordinates	9a	
		9b	
		9c	
10	Focus on algebra	10a	
		10b	
11	Solving more problems	11a	
		11b	
		11c	
12	Fractions, equivalents and algebra	12a	
		12b	
		12c	
13	Fair shares	13a	
		13b	
		13c	
14	Nets, angles and coordinates	14a	
		14b	
		14c	

RISING STARS
Mathematics

Year
6

Practice Book

Name:

Contents

Whole and part numbers

1a   Place value

1   Write these as numbers.

a   six hundred and twenty-eight thousand

> [   ]

b   one hundred and seventy-six thousand six hundred and thirty-five

> [   ]

c   two hundred and ninety-seven thousand three hundred and ninety

> [   ]

d   three hundred and four thousand

> [   ]

e   eight hundred thousand seven hundred and twelve

> [   ]

f   five million nine hundred and fifteen thousand

> [   ]

g   four million one hundred and sixty-seven thousand eight hundred and fifteen

> [   ]

h   two million five hundred and eighteen thousand eight hundred and thirty

> [   ]

i   nine million two hundred and thirty-seven thousand one hundred and ten

> [   ]

j   seven million eight hundred thousand nine hundred and six

> [   ]

2 Circle the digit that matches the value.

a 483 385 three thousand

b 140 441 forty

c 299 600 ninety thousand

d 557 725 five hundred thousand

e 806 636 six

f 1 922 505 two thousand

g 3 380 788 eighty

h 7 677 973 seven million

i 5 444 129 four hundred thousand

j 1 911 048 ten thousand

 3 Write the value of each digit in these numbers.

$$245.67 \rightarrow 200 + 40 + 5 + \frac{6}{10} + \frac{7}{100}$$

a 432.51

b 165.34

c 4351.26

d 2514.63

e 45 143.52

f 253 626.11

 4 Answer these.

a $256 \times 10 =$

b $1748 \times 100 =$

c $3.9 \times 100 =$

d $45.29 \times 10 =$

e $81 893 \times 100 =$

f $621 047 \times 10 =$

g $95 431.52 \times 10 =$

h $21 406.78 \times 100 =$

5 Here is part of a computer spreadsheet.

	Input	A	B	C	D	E
1	36	360	3.6	3600	0.36	36 000
2	37	370	3.7	3700	0.37	37 000
3	38	380	3.8	3800	0.38	38 000
4	39					
5	40					
6	41					
7	42					
8	43					

a What does column A do to the input number?

b What does column B do to the input number?

c What does column C do to the input number?

d What does column D do to the input number?

e What does column E do to the input number?

f Write the missing numbers in the spreadsheet.

g If the input was 105, what would be in column D?

h If the input was 17.5, what would be in column E?

i If 82 was at position 9B, what would be at 9E?

j If 1.25 was at position 10D, what would be at 10A?

k If 16 550 was at position 11C, what would be at 11B?

6

6   A scientist measured different plants in metres. Complete the chart.

Plant	Height (m)	Height (cm)	Height (mm)
Fern	2.7 m		
Mushroom	0.08 m		
Redwood tree	92.6 m		
Lily flower	0.63 m		
Rhododendron bush	3.95 m		
Pine tree	10.24 m		

 7

YOU WILL NEED:
- calculator
- partner

Use a non-scientific calculator to investigate decimal patterns.

Key this in to make a ×100 machine:

Repeat this, changing 0.006 to other numbers.
Explain what is happening to a partner.

1 Circle the **largest** number in each pair.

a   11 354   11 345   f   721 734   712 374

b   80 296   80 469   g   253 309   253 009

c   39 457   34 857   h   104 739   14 930

d   612 203   212 326   i   95 876   859 967

e   985 686   985 689   j   655 643   665 634

2 Complete the table with these numbers so that they are in order.

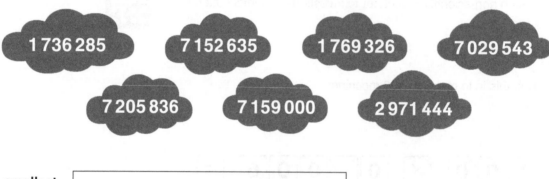

1 736 285   7 152 635   1 769 326   7 029 543

7 205 836   7 159 000   2 971 444

smallest	
largest	

3 Round these numbers to the nearest whole number. Then round them to the nearest 10 and the nearest 100.

	Nearest whole number	Nearest 10	Nearest 100
945.2			
1884.7			
239.6			
1940.3			
667.15			
9080.21			
330.04			
7425.99			

4 Round each amount to the nearest pound to work out the **approximate** answers.

a The prices of 2 books are £9.80 and £6.30. What is the approximate cost if they are bought together?

£

b A man spent £29.50 on a shirt and £12.40 on a tie. How much did he spend approximately?

£

c A cricket bat costs £53.90 and a ball costs £8.20. What is the approximate total price of the bat and ball?

£

d Jade bought a pizza for £7.85 and then had a bowl of ice cream for £3.75. How much did she spend on this meal approximately?

£

e A table costs £56.49 and a chair costs £21.65. What is the approximate price of the table and chair together?

£

f Freddie bought a phone for £64.99 and a cover for £6.72. What was the approximate total price of the phone and cover?

£

Move a counter or coin horizontally or vertically. You can only move one place at a time and you must move to a number that is **smaller** than the one you are on.

a What is the shortest route you can find from START to FINISH? Record the numbers as you go or use coloured pencils.

b Can you find 5 different routes?

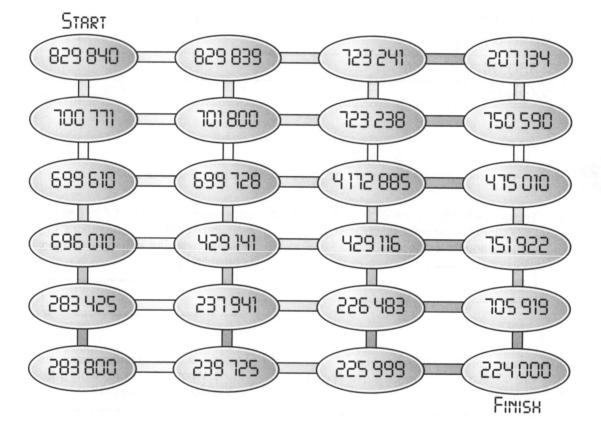

Start

829 840	829 839	723 241	207 134
700 771	701 800	723 238	750 590
699 610	699 728	4 172 885	475 010
696 010	429 141	429 116	751 922
283 425	237 941	226 483	705 919
283 800	239 725	225 999	224 000

Finish

6 Investigate the areas of the some of the largest deserts in the world.
Complete this table to show your findings. Put the deserts in order, starting with the **largest**.

Name of desert	Area (km²)	Area to the nearest 100 km²

Use the fraction wall to help you answer these questions.

1											
$\frac{1}{2}$						$\frac{1}{2}$					
$\frac{1}{3}$				$\frac{1}{3}$				$\frac{1}{3}$			
$\frac{1}{4}$			$\frac{1}{4}$			$\frac{1}{4}$			$\frac{1}{4}$		
$\frac{1}{5}$		$\frac{1}{5}$		$\frac{1}{5}$		$\frac{1}{5}$			$\frac{1}{5}$		
$\frac{1}{6}$		$\frac{1}{6}$		$\frac{1}{6}$		$\frac{1}{6}$		$\frac{1}{6}$		$\frac{1}{6}$	
$\frac{1}{8}$	$\frac{1}{8}$	$\frac{1}{8}$	$\frac{1}{8}$		$\frac{1}{8}$	$\frac{1}{8}$	$\frac{1}{8}$		$\frac{1}{8}$		
$\frac{1}{10}$	$\frac{1}{10}$	$\frac{1}{10}$	$\frac{1}{10}$	$\frac{1}{10}$	$\frac{1}{10}$	$\frac{1}{10}$	$\frac{1}{10}$	$\frac{1}{10}$	$\frac{1}{10}$		
$\frac{1}{12}$	$\frac{1}{12}$	$\frac{1}{12}$	$\frac{1}{12}$	$\frac{1}{12}$	$\frac{1}{12}$	$\frac{1}{12}$	$\frac{1}{12}$	$\frac{1}{12}$	$\frac{1}{12}$	$\frac{1}{12}$	$\frac{1}{12}$

1 Use the symbols <, > or = to compare these fractions.

a $\frac{1}{2}$ ☐ $\frac{3}{10}$

b $\frac{1}{4}$ ☐ $\frac{2}{10}$

c $\frac{3}{5}$ ☐ $\frac{3}{4}$

d $\frac{6}{12}$ ☐ $\frac{1}{2}$

e $\frac{1}{3}$ ☐ $\frac{3}{8}$

f $\frac{1}{10}$ ☐ $\frac{2}{5}$

g $\frac{2}{3}$ ☐ $\frac{6}{10}$

h $\frac{5}{6}$ ☐ $\frac{11}{12}$

i $\frac{4}{6}$ ☐ $\frac{1}{2}$

j $\frac{6}{8}$ ☐ $\frac{1}{4}$

2 Circle the **largest** fraction in each set.

a
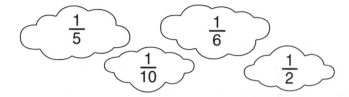

b
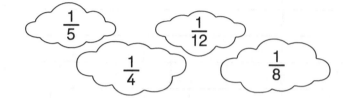

c

d
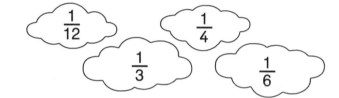

e
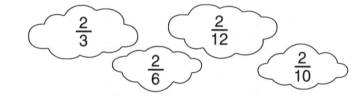

f

g
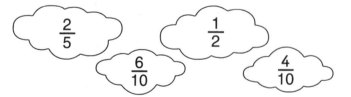

h
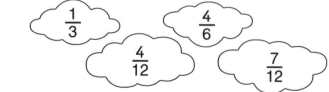

3 Complete these equivalent fraction chains.

a $\dfrac{1}{4}$ = $\dfrac{2}{8}$ = $\dfrac{3}{\boxed{}}$ = $\dfrac{4}{\boxed{}}$ = $\dfrac{5}{\boxed{}}$

b $\dfrac{2}{3}$ = $\dfrac{4}{6}$ = $\dfrac{6}{\boxed{}}$ = $\dfrac{8}{\boxed{}}$ = $\dfrac{10}{\boxed{}}$

c $\dfrac{3}{5}$ = $\dfrac{6}{\boxed{}}$ = $\dfrac{9}{\boxed{}}$ = $\dfrac{12}{\boxed{}}$ = $\dfrac{15}{\boxed{}}$

> What do you notice about the patterns?

d $\dfrac{5}{8}$ = $\dfrac{10}{\boxed{}}$ = $\dfrac{15}{\boxed{}}$ = $\dfrac{20}{\boxed{}}$ = $\dfrac{25}{\boxed{}}$

 4 Simplify these fractions. Show the common factor you divided by.

$$\dfrac{12}{18} = \boxed{\dfrac{2}{3}} \quad \boxed{\div\ 6}$$

a $\dfrac{4}{20}$ = $\boxed{}$ $\boxed{\div\ }$

b $\dfrac{5}{35}$ = $\boxed{}$ $\boxed{\div\ }$

c $\dfrac{10}{12}$ = $\boxed{}$ $\boxed{\div\ }$

d $\dfrac{6}{9}$ = $\boxed{}$ $\boxed{\div\ }$

e $\dfrac{12}{16}$ = $\boxed{}$ $\boxed{\div\ }$

f $\dfrac{9}{15}$ = $\boxed{}$ $\boxed{\div\ }$

g $\dfrac{25}{30}$ = $\boxed{}$ $\boxed{\div\ }$

h $\dfrac{14}{21}$ = $\boxed{}$ $\boxed{\div\ }$

5 Dan knew straight away that $\frac{10}{26}$ cannot be equivalent to $\frac{2}{5}$ because 26 is not a multiple of 5.

Cross out the fractions that **cannot be** equivalent to $\frac{2}{5}$.

Circle the fractions that **could be** equivalent to $\frac{2}{5}$.

Tick the fractions that **are** equivalent to $\frac{2}{5}$.

$\frac{15}{32}$	$\frac{12}{25}$	$\frac{18}{45}$
$\frac{15}{40}$	$\frac{32}{48}$	$\frac{17}{54}$
$\frac{12}{30}$	$\frac{30}{75}$	$\frac{20}{52}$

6

YOU WILL NEED:
* **number cards 6–16**
* **partner**

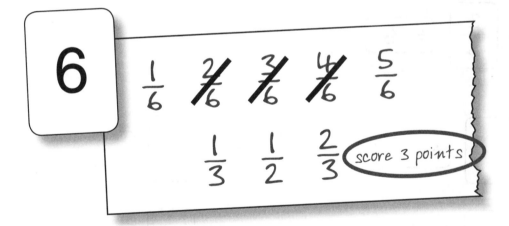

* Shuffle the number cards and place them face down in a pile.

* Player 1 takes the top card. Player 1 writes the set of fractions less than 1 using the number on the card as the denominator. This player changes any that can be simplified and scores a point for each of these.

* Player 2 takes a turn.

1 Write the equivalent fraction with a denominator of 30 for each fraction below.

$$\frac{1}{2} = \boxed{\frac{15}{30}}$$

a $\frac{1}{3} = \boxed{}$

d $\frac{1}{10} = \boxed{}$

g $\frac{5}{6} = \boxed{}$

b $\frac{1}{5} = \boxed{}$

e $\frac{2}{3} = \boxed{}$

h $\frac{7}{10} = \boxed{}$

c $\frac{1}{6} = \boxed{}$

f $\frac{3}{5} = \boxed{}$

i $\frac{2}{5} = \boxed{}$

2

a Write your answers from question 1 on this number line.

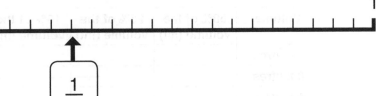

$\boxed{\frac{15}{30}}$

0 ————————————————————— 1

$\boxed{\frac{1}{2}}$

b Which is the **largest** fraction?

c Which is the **smallest** fraction?

3 Complete these equivalent fractions.

a $\dfrac{2}{9} = \dfrac{\boxed{}}{18}$

e $\dfrac{10}{15} = \dfrac{\boxed{}}{3}$

i $\dfrac{20}{24} = \dfrac{5}{\boxed{}}$

b $\dfrac{4}{5} = \dfrac{8}{\boxed{}}$

f $\dfrac{1}{3} = \dfrac{\boxed{}}{12}$

j $\dfrac{2}{5} = \dfrac{14}{\boxed{}}$

c $\dfrac{1}{2} = \dfrac{\boxed{}}{14}$

g $\dfrac{12}{16} = \dfrac{3}{\boxed{}}$

k $\dfrac{15}{24} = \dfrac{\boxed{}}{8}$

d $\dfrac{7}{28} = \dfrac{1}{\boxed{}}$

h $\dfrac{3}{8} = \dfrac{\boxed{}}{16}$

l $\dfrac{4}{7} = \dfrac{12}{\boxed{}}$

4 Complete this equivalents grid.

Fractions	Percentages	Decimals
$\frac{1}{10}$		
	30%	
		0.5
$\frac{2}{10}$		
	60%	
		0.4
$\frac{9}{10}$		
	70%	
		0.8

5

a Complete this chart, writing each volume in millilitres.

Volume	50% of the volume (ml)	25% of the volume (ml)	10% of the volume (ml)	20% of the volume (ml)	5% of the volume (ml)	1% of the volume (ml)
3.6 litres						
8.2 litres						
4.8 litres						
16.4 litres						
20.2 litres						
24.6 litres						

b Explain how you could use the table above to work out 36% of 3.6 litres.

_____

_____

6 Look at this shape puzzle.

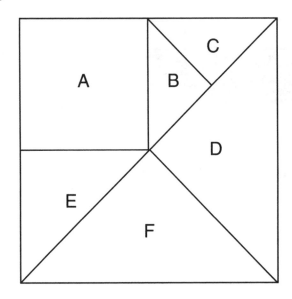

a What fraction of the large square is each shape?

A →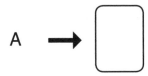

C →

E →

B →

D →

F →

b Add each of your fractions together. What is the total?

c The area of the large square is 25 cm².

What is the area of the small square?

Calculations and algebra

1 Use rounding and adjusting to answer these. Show your working.

2508 + 3495 − 199 = $\boxed{5804}$

Working:

2500 + 3500 − 200 → 6000 − 200 = 5800,
then add 8 and subtract 5 and add 1

a 509 + 398 − 201 =

Working:

b 498 + 704 − 195 =

Working:

c   2302 + 497 − 603 = ☐

Working:

d   5498 + 805 − 1999 = ☐

Working:

e   1601 + 4199 − 503 = ☐

Working:

f   3797 + 2603 − 198 = ☐

Working:

2 Use counting on to answer these.

$4904 - 4397 =$ | 507 |

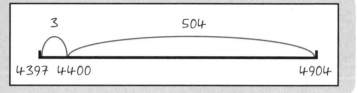

a $3905 - 3148 =$ | |

b $8762 - 8293 =$ | |

c $8472 - 6800 =$ | |

d $6036 - 2449 =$ | |

e $8165 - 1295 =$ | |

f $9962 - 6355 =$ | |

3 Use the sequencing strategy to add these lengths. Show how you partition the smaller number.

> 5748 mm + 3124 mm = $\boxed{8872}$ mm
>
> Working:
>
> 5748 + 3000 + 100 + 20 + 4

a 2507 mm + 3340 mm = ☐ mm

Working:

b 5629 mm + 1257 mm = ☐ mm

Working:

c 1064 mm + 3299 mm = ☐ mm

Working:

d   2206 mm + 2977 mm = [___] mm

Working:

e   4893 mm + 3307 mm = [___] mm

Working:

f   7215 mm + 1885 mm = [___] mm

Working:

4 Use the prices below to answer these.

violin
£185.20

trumpet
£208.50

keyboard
£399.50

banjo
£424.99

drum kit
£549.95

harp
£370.89

a What does it cost to buy a harp and a trumpet?

£ _____

b A school buys 2 violins. What is the total price?

£ _____

c How much will you spend if you buy a keyboard and a drum kit?

£ _____

d How much more will it cost to buy a trumpet than a violin?

£ _____

e What is the difference in price between a harp and a banjo?

£ _____

f Which 2 instruments can you buy for exactly £633.49?

_____ and _____

5 Answer these problems using your own mental strategies.

a The Olympic games were held in London in 1948 and 2012. How many years after the 1948 games was it before London hosted the Olympic games again?

d A supermarket sells 3093 bread rolls on Friday and 2160 bread rolls on Saturday. How many bread rolls are sold in total over these 2 days?

b A building site has 2 deliveries of bricks. The first lorry brings 4598 bricks and the next lorry delivers 3015 bricks. How many bricks are delivered altogether?

e The walls on a new house are 6670 mm high. When the roof is on, the house will be 8349 mm high. What is the height of the roof?

mm

c A holiday for a family costs £4275 in August and £2850 in January. How much cheaper is it to go on holiday in January than August?

£

f The distance to fly from London to Rome is 1489 km. How far will a plane fly on a return flight between London and Rome?

km

Which mental strategies did you prefer to use to answer these?

6 There are 5 divers exploring a wreck. Each has an oxygen tank that can hold up to 10 000 units of oxygen. 1000 units of oxygen lasts 1 minute.

a Complete this chart to show how many units are left in each tank.

Diver	Oxygen used	Oxygen left
Amy	7600 units	
Ben	4894 units	
Clare	6702 units	
Dhruv	5987 units	
Evie	6413 units	

b Using the amount of oxygen left in each tank, which **3** tanks would make exactly 1 full tank of 10 000 units?

_____

Bidmas table

1st	**B**rackets
2nd	**I**ndices e.g. $(5 \times 2)^2$
3rd	**D**ivision & **M**ultiplication
Last	**A**ddition & **S**ubtraction

1 Answer these. Remember to use the correct order of operations.

a $3746 + 8^2 =$ ⬚

b $195 + 2045 - 832 =$ ⬚

c $25 \times 12 + 677 =$ ⬚

d $8349 - 420 \div 6 =$ ⬚

e $57 + 93 \times 8 =$ ⬚

f $480 \div (\frac{1}{4} \times 24) =$ ⬚

g $7 \times (-162 \div 3) =$ ⬚

h $(9^2 + 63) \div 12 - 5 =$ ⬚

2 Use <, > or = to make each of these true.

a $7 + 11^2 - 60$ ⬚ $15 + (4 \times 12)$

b $38 - (72 \div 12)$ ⬚ $90 - 7 \times 8$

c $(16 + 9) \div 5$ ⬚ $50 - 15 \times 3$

d $(13 - 5) + (8 \times 15)$ ⬚ $68 + (4 \times 3)^2$

e $7 \times 6^2 \div 9$ ⬚ $(21 \times 4) \div 3$

f $56 + (10 \times 2.4)$ ⬚ $(30 \times 8) \div 4$

g $372 + 8^2 - 119$ ⬚ $153 + (14 \times 12)$

h $(808 - 55) \div 3$ ⬚ $228 + (12 \times 2)$

3 Put brackets in these to make the largest possible answer.

a $7 + 9 \times 6 =$ ____

b $96 - 56 \div 8 =$ ____

c $12^2 \div 6 \times \frac{1}{2} =$ ____

d $90 \div 5 + 4 \times 3 =$ ____

e $26 - 2 \times 8 + 20 =$ ____

f $2 \times 35 + 560 \div 7 =$ ____

4 Answer these. Show your working.

a There are 3 printers. Each weighs 13.5 kg. They are put in boxes that each weigh 1.4 kg and then put in a crate that weighs 7.6 kg. What is the total weight of the full crate?

____ kg

Working:

c A recipe for a fruit smoothie uses 360 ml of raspberries, 174 ml bananas and 216 ml of milk. Sam wants to make 1.5 litres of smoothie. How much milk will he need?

____ ml

Working:

b The distance between the bus station and the school is 18 km. The bus takes this journey to the school and back to the bus station every day in the morning and in the evening. How far has the bus travelled in 1 week from Monday to Friday?

____ km

Working:

d Wooden floorboards come in 3 widths: 30 cm, 80 cm and 120 cm. The width of a room is 5.5 m. How many floorboards of each size will fit the width of the floor exactly without having to cut any?

____

____

____

Working:

5 Play a game of target numbers.

Use any of the 7 numbers given to get as near as possible to the target number.

You can only use each number once.

You can use brackets and the operations +, −, × and ÷.

a

2010 = _____

b

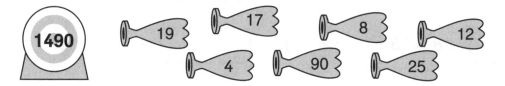

1490 = _____

c

2655 = _____

d

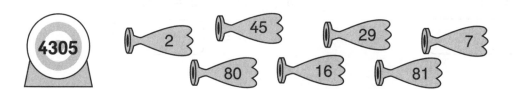

4305 = _____

e

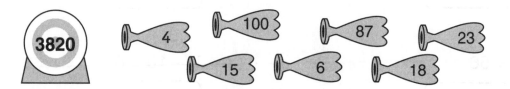

3820 = _____

Make up your own target numbers and play the game.

1 Write the missing numbers.

a   27 + [   ] = 52

b   [   ] × 8 = 96

c   3015 − [   ] = 2994

d   [   ] ÷ 8 = 15

e   [   ] − 463 = 147

f   7 × [   ] = 630

g   882 + [   ] = 1000

h   200 ÷ [   ] = 5

i   [   ] + 6.2 = 9

j   [   ] − 5.4 = 1.6

k   3 × [   ] = 84

l   [   ] ÷ 12 = 30

 2 Write the value of each letter.

a $c + 8 = 19$

     $c =$ [   ]

b $37 - y = 19$

     $y =$ [   ]

c $7 \times b = 56$

     $b =$ [   ]

d $48 \div z = 8$

     $z =$ [   ]

e $h - 1.5 = 2.5$

     $h =$ [   ]

f $95 + a = 104$

     $a =$ [   ]

g $(d \times 3) - 10 = 8$

     $d =$ [   ]

h $6r = 24$

     $r =$ [   ]

i $(4 + s) \times 10 = 110$

$s =$ [ ]

k $28 - 2f = 10$

$f =$ [ ]

j $5 \times (9 - e) = 35$

$e =$ [ ]

l $60 \div (4 + t) = 5$

$t =$ [ ]

 3 Use number rods as shown.

Write 2 different algebraic expressions for each.

a

	c	
d	d	e

[ ]

[ ]

c

a	a
b	c

[ ]

[ ]

b

	v	
w		x

[ ]

[ ]

d

n	n
m	m	p

[ ]

[ ]

 4 Rod z is 60. The difference between w and x is 36.

What is the value of rods w and x?

	z	
w	x	

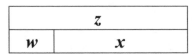

Rod w is [ ]

Rod x is [ ]

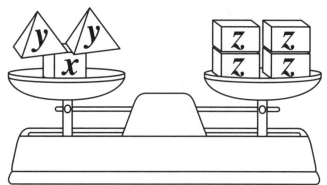

a Write an algebraic expression for this.

_____

_____

_____

b Work out the mass of each object x, y and z from this information:

$x + y = 310\,g$

$x - y = 50\,g$

$x = \boxed{\ g}$

$y = \boxed{\ g}$

$z = \boxed{\ g}$

3a   Using long multiplication

⭐ **1**   Complete these multiplication super-squares.

×2 →

	×2		
	2	4	8
×7 ↓	14	**28**	56
	98	196	392

a

×8 →

×3 ↓		**24**	

c

×4 →

×11 ↓		**132**	

b

×5 →

×9 ↓		**90**	

d   Make up your own super-square.

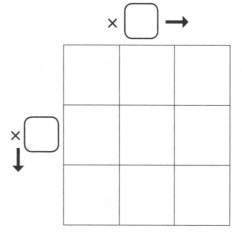

× ▢ →

× ▢ ↓

Use the grid method to answer these.

a   34 × 56 = [   ]

d   53 × 64 = [   ]

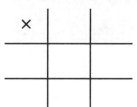

b   63 × 45 = [   ]

e   54 × 63 = [   ]

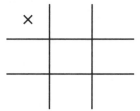

c   43 × 56 = [   ]

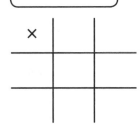

3

Use the grid method to answer these.

a   127 × 29 = [   ]

c   596 × 39 = [   ]

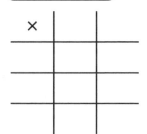

b   394 × 42 = [   ]

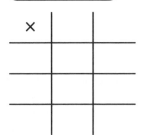

d   749 × 55 = [   ]

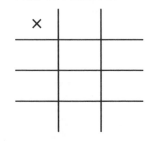

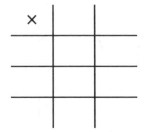

4 Answer these using the long multiplication method.

a 381 × 24

d 608 × 73

b 420 × 57

e 234 × 96

c 816 × 18

Compare your working out for questions 3 and 4. Which do you prefer?

 5 Solve these.

a There are 24 hours in a day. How many hours are there in January?

d There are 18 bottles of water in a multi-pack and each bottle holds 350 ml. How much water is there altogether?

 ml

b A school has 18 classes each with 32 children. How many children are there altogether at this school?

e A farm has 177 sacks of potatoes, each with mass 26 kg. What is the total mass of these potatoes?

kg

c 47 buses arrive at a theme park on 1 day. Each bus is full and has 54 seats. How many visitors in total are going to the theme park?

f A new film is so popular that a cinema is full every time the film is shown. Over 1 week, the film is shown 36 times to an audience of 289 people. How many people watched the film in this 1 week?

6 This is the 'Russian Peasant Method' for multiplying large numbers. It only involves halving, doubling and adding.

Follow this example for 41 × 74.

• Write each number at the top of a column.
• Keep halving the number in the left column until you reach 1. If it is an odd number, take away 1 and halve that number instead.

• Double the numbers in the right column for the same number of steps as in the left column.

• Cross out any even numbers in the left column and the corresponding numbers in the right column.

• Add the numbers in the right column that are not crossed out. This gives the answer to the multiplication.

41	74
~~20~~	~~148~~
~~10~~	~~296~~
5	592
~~2~~	~~1184~~
1	2368
	3034

Try this method out, multiplying with your own numbers. Which do you prefer – the 'Russian Peasant Method' or the 'Gelosia Method' (see page 49 in the Textbook)?

⭐ **1**   Complete these multiplication chains.

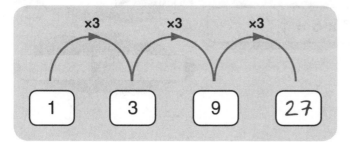

×3   ×3   ×3

| 1 | 3 | 9 | 27 |

a

×2   ×2   ×2   ×2   ×2

| 1 | 2 | 4 | | | |

b

×3   ×3   ×3   ×3   ×3

| 1 | | | | | |

c

×4   ×4   ×4   ×4   ×4

| 1 | | | | | |

d

×5   ×5   ×5   ×5   ×5

| 1 | | | | | |

What do you notice about the patterns?

2 Use each fact to help answer the others. Make up another fact for each question.

a

700 × 60 = ☐

70 × 60 = ☐

7 × 6 = ☐

17 × 6 = ☐

107 × 6 = ☐

☐ = ☐

b

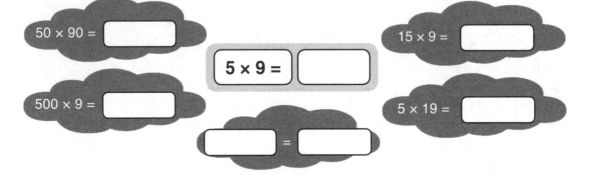

50 × 90 = ☐

15 × 9 = ☐

5 × 9 = ☐

500 × 9 = ☐

5 × 19 = ☐

☐ = ☐

c

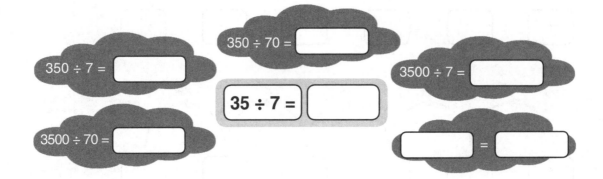

350 ÷ 70 = ☐

350 ÷ 7 = ☐

3500 ÷ 7 = ☐

35 ÷ 7 = ☐

3500 ÷ 70 = ☐

☐ = ☐

d

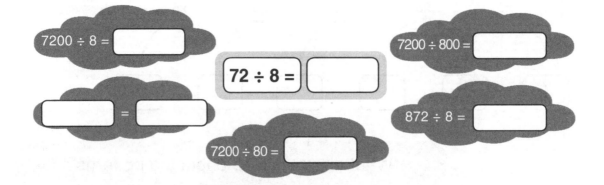

7200 ÷ 8 = ☐

7200 ÷ 800 = ☐

72 ÷ 8 = ☐

☐ = ☐

872 ÷ 8 = ☐

7200 ÷ 80 = ☐

3 Answer these using mental methods.

Describe your method for each question.

a   38 × 5 = ☐

b   120 × 6 = ☐

c   231 × 4 = ☐

d   700 × 12 = ☐

e   82 × 50 = ☐

f   800 × 60 = ☐

_____

_____

4 Answer these using mental methods.

Describe your method for each question.

a   180 ÷ 9 = ☐

b   340 ÷ 2 = ☐

c   250 ÷ 50 = ☐

d   816 ÷ 4 = ☐

e   900 ÷ 30 = ☐

f   4200 ÷ 6 = ☐

_____

_____

5 Solve these.

a   In a library there are 7 shelves of children's books and each shelf has 90 books. How many children's books are there altogether?

☐

b   To drive from Liverpool to London and back again is 600 km. A lorry does this distance every day from Monday to Friday. How far does the lorry travel in a week?

 km

c   There are 12 coloured crayons in a pack. A school buys 40 packs. How many crayons are there in total?

☐

d   Kate buys 2 identical sofas for £320. What was the price of 1 sofa?

 £

e A kitchen is 270 cm wide and floor tiles are 30 cm in length. How many tiles will fit across the width of this kitchen?

f A ski holiday for 4 people costs a total of £2800. How much will it cost for each person to go on this holiday?

£

6 This is a multiplication arithmagon.

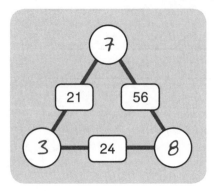

a

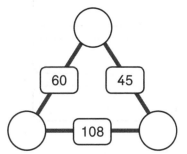

c

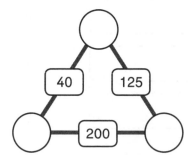

b

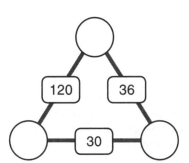

d

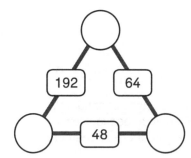

1 Use the fact to calculate the others. Look at the patterns in the answers.

a

0.7 × 4 = ☐

7 × 4 = ☐   7 × 0.4 = ☐

0.7 × 0.4 = ☐

b

8 × 0.9 = ☐

8 × 9 = ☐   0.8 × 9 = ☐

0.8 × 0.9 = ☐

c

5.6 ÷ 7 = ☐

56 ÷ 7 = ☐   0.56 ÷ 0.7 = ☐

0.56 ÷ 7 = ☐

d

0.54 ÷ 0.6 = ☐

54 ÷ 6 = ☐   0.54 ÷ 6 = ☐

5.4 ÷ 6 = ☐

2 Complete these multiplication squares.

a

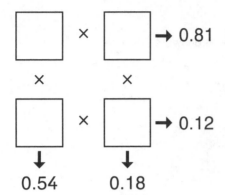

□ × □ → 0.81

× ×

□ × □ → 0.12

↓ ↓

0.54 0.18

c

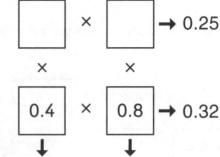

□ × □ → 0.25

× ×

0.4 × 0.8 → 0.32

↓ ↓

b

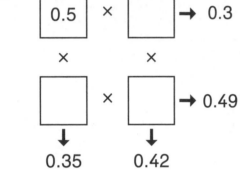

0.5 × □ → 0.3

× ×

□ × □ → 0.49

↓ ↓

0.35 0.42

Try to make up your own decimal multiplication square like these.

3 Work out the missing numbers.

a 0.7 × □ = 3.5

b □ ÷ 5 = 0.4

c 1.2 ÷ □ = 3

d □ × 0.9 = 0.27

e □ ÷ 0.4 = 8

f 0.2 × □ = 0.18

g 0.49 ÷ □ = 0.7

h 2.4 ÷ □ = 0.4

4 Solve these problems.

a Cinema tickets cost £3.80. What is the total cost of 6 tickets?

£ [   ]

b The heights of 5 players in a school basketball team are 1.53 m, 1.5 m, 1.56 m, 1.57 m and 1.49 m. What is their mean height?

[   ] m

c There is 1.8 litres of juice in a jug and it is poured equally into 6 glasses. How much juice is in each glass?

[   ] litres

d Melons cost £1.47 each, but there is a 'Buy 3 for the price of 2 offer'. What is the individual price of each melon in this offer?

[   ] p

e A baker makes 15 kg of cake mixture to make 6 equal mass wedding cakes. Each cake is sliced into 50 equal slices. What is the mass of each slice of cake?

[   ] kg

 5 Use 4 different digits to complete this so that the answer is a 1-place decimal number.

Find 5 different ways to do this.

| 1 | 2 | 3 | 4 | 5 |

| 6 | 7 | 8 | 9 |

 ⬜⬜.⬜ ÷ ⬜

 ⬜⬜.⬜ ÷ ⬜

 ⬜⬜.⬜ ÷ ⬜

 ⬜⬜.⬜ ÷ ⬜

⬜⬜.⬜ ÷ ⬜

1

Write the value of the black rod if the grey rod is:

a   6 → [   ]   d   90 → [   ]

b   8 → [   ]   e   3.5 → [   ]

c   15 → [   ]   f   0.02 → [   ]

2

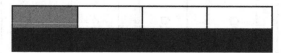

Write the value of the grey rod if the black rod is:

a   24 → [   ]   d   9.6 → [   ]

b   30 → [   ]   e   0.4 → [   ]

c   104 → [   ]   f   3.8 → [   ]

3 Answer these.

a $\frac{1}{4}$ of 16 = [ ]

   $\frac{3}{4}$ of 16 = [ ]

b $\frac{1}{3}$ of 21 = [ ]

   $\frac{2}{3}$ of 21 = [ ]

c $\frac{1}{6}$ of 18 = [ ]

   $\frac{5}{6}$ of 18 = [ ]

d $\frac{1}{5}$ of 60 = [ ]

   $\frac{4}{5}$ of 60 = [ ]

e $\frac{1}{7}$ of 21 = [ ]

   $\frac{4}{7}$ of 21 = [ ]

f $\frac{1}{8}$ of 40 = [ ]

   $\frac{3}{8}$ of 40 = [ ]

g $\frac{1}{10}$ of 200 = [ ]

   $\frac{7}{10}$ of 200 = [ ]

h $\frac{1}{10}$ of 14 = [ ]

   $\frac{2}{10}$ of 14 = [ ]

 4 What are the mystery numbers?

a I am thinking of a number. 45 is 9 times larger than my number. What is my number?

   [ ]

b I am thinking of a number. When I divide it by 7 my answer is 6. What is my number?

   [ ]

c I am thinking of a number. If I add 4 and then multiply it by 5 my answer is 60. What is my number?

   [ ]

d I am thinking of a number. If I divide it by 9 and then subtract 2 the answer is 6. What is my number?

   [ ]

e I am thinking of two numbers. One of my numbers is $\frac{2}{5}$ of the other. One of my numbers is 20. What are the possible values of the other number?

   [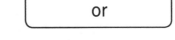 or]

f I am thinking of a number. One of my numbers is 24. One of my numbers is $\frac{3}{4}$ of the other number. What are the possible values of the other number?

   [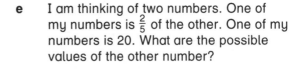 or]

There are 14 children's programmes on Channel CTV each weekday.

This pie chart show the proportions for each type of programme.

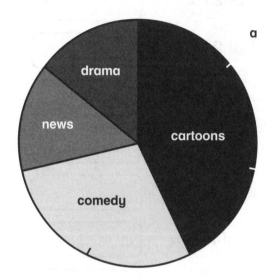

a What is the fraction of each type of programme?

news → ⬜   drama → ⬜

cartoons → ⬜   comedy → ⬜

b How many cartoon programmes are there each day?

⬜

c How many comedy programmes are there each day?

⬜

d Use this number line to find out the total number of drama programmes from Monday to Friday.

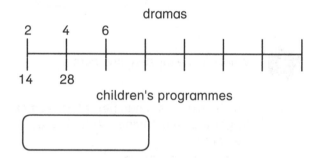

dramas

2   4   6

14   28

children's programmes

⬜

e Now do the same for cartoon programmes.

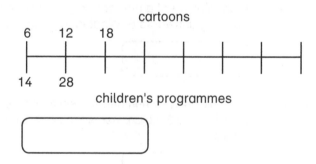

cartoons

6   12   18

14   28

children's programmes

⬜

2-D shapes, 3-D shapes and nets

1 Calculate the area and perimeter of these rectangles.

a area = ___ cm²

perimeter = ___ cm

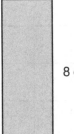

8 cm

3 cm

d area = ___ cm²

perimeter = ___ cm

2 cm

18 cm

b area = ___ cm²

perimeter = ___ cm

4 cm

6 cm

e area = ___ cm²

perimeter = ___ cm

6 cm

6 cm

c area = ___ cm²

perimeter = ___ cm

12 cm

2 cm

f area = ___ cm²

perimeter = ___ cm

9 cm

4 cm

Draw a diagonal on these rectangles and shade one of the triangles that you make.

Calculate the area of each rectangle and then each triangle.

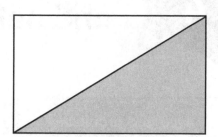

Each square is 1 cm²

a

c

b

d

a   area of rectangle = ⬚ cm²

area of triangle = ⬚ cm²

d   area of rectangle = ⬚ cm²

area of triangle = ⬚ cm²

b   area of rectangle = ⬚ cm²

area of triangle = ⬚ cm²

e   Now describe how you can calculate the area of a right-angled triangle.

c   area of rectangle = ⬚ cm²

area of triangle = ⬚ cm²

Use the method you described in question 2 to calculate the area of these right-angled triangles.

a   area of triangle =

$\boxed{\qquad\text{m}^2}$

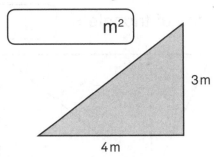

3m

4m

d   area of triangle =

$\boxed{\qquad\text{m}^2}$

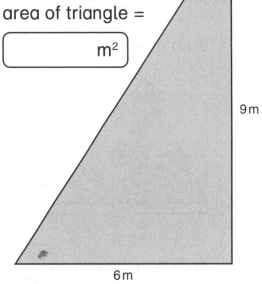

9m

6m

b   area of triangle =

$\boxed{\qquad\text{m}^2}$

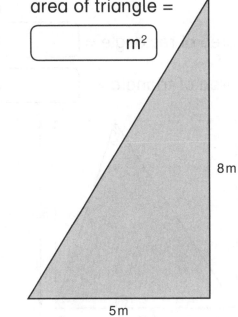

8m

5m

e   area of triangle =

$\boxed{\qquad\text{m}^2}$

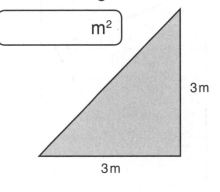

3m

3m

c   area of triangle =

$\boxed{\qquad\text{m}^2}$

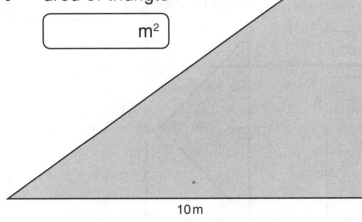

7m

10m

f   area of triangle =

$\boxed{\qquad\text{m}^2}$

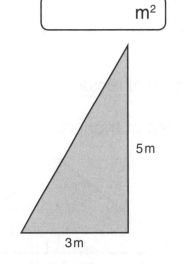

5m

3m

4 Calculate the area of the shaded triangle. How does the rectangle help you?

a area of rectangle = [　　　] cm²

area of triangle = [　　　] cm²

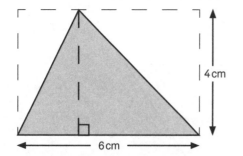

4 cm

6 cm

d area of rectangle = [　　　] cm²

area of triangle = [　　　] cm²

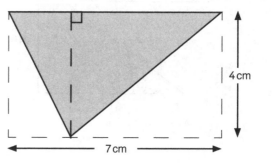

4 cm

7 cm

b area of rectangle = [　　　] cm²

area of triangle = [　　　] cm²

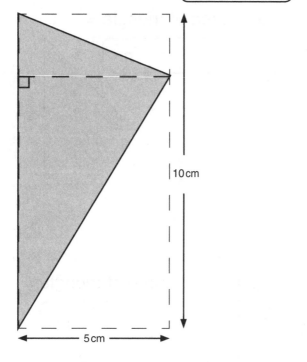

10 cm

5 cm

e area of rectangle = [　　　] cm²

area of triangle = [　　　] cm²

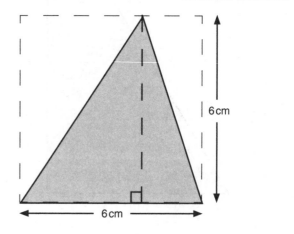

6 cm

6 cm

f area of rectangle = [　　　] cm²

area of triangle = [　　　] cm²

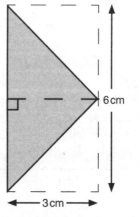

6 cm

3 cm

c area of rectangle = [　　　] cm²

area of triangle = [　　　] cm²

3 cm

8 cm

5 Calculate the area of these parallelograms.

a area of
parallelogram = [ ] cm²

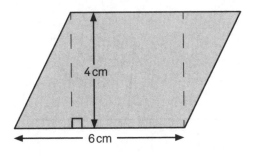

b area of
parallelogram = [ ] cm²

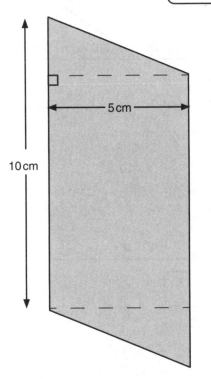

c area of
parallelogram = [ ] cm²

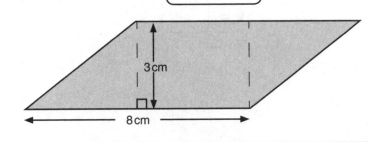

d area of
parallelogram = [ ] cm²

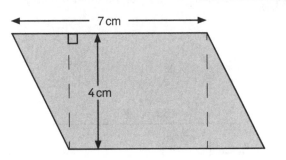

e area of
parallelogram = [ ] cm²

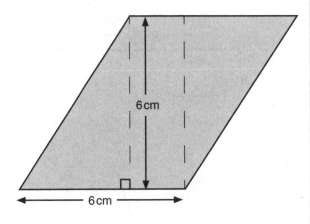

f area of
parallelogram = [ ] cm²

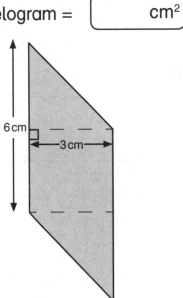

1 Calculate the size of the unknown angles.

a

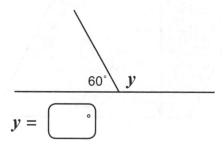

60° *y*

$y = $ ◯ °

b

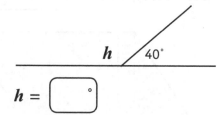

h 40°

$h = $ ◯ °

c

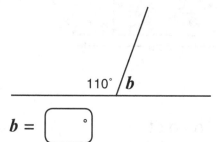

110° *b*

$b = $ ◯ °

d

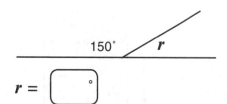

150° *r*

$r = $ ◯ °

e

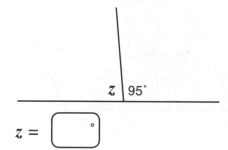

z 95°

$z = $ ◯ °

f

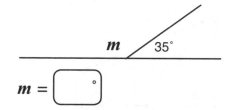

m 35°

$m = $ ◯ °

g

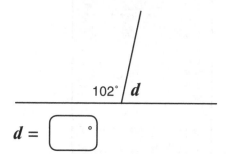

102° *d*

$d = $ ◯ °

h

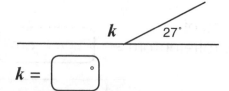

k 27°

$k = $ ◯ °

2 Calculate the size of the unknown angles.

a

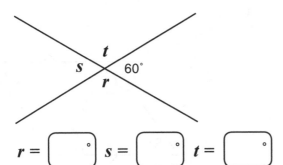

$r =$ ⬚ ° $s =$ ⬚ ° $t =$ ⬚ °

d

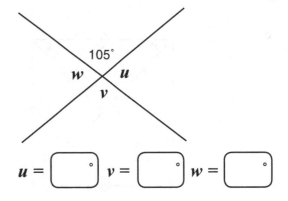

$u =$ ⬚ ° $v =$ ⬚ ° $w =$ ⬚ °

b

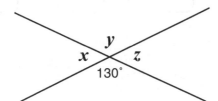

$x =$ ⬚ ° $y =$ ⬚ ° $z =$ ⬚ °

e

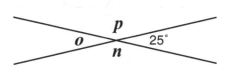

$n =$ ⬚ ° $o =$ ⬚ ° $p =$ ⬚ °

c

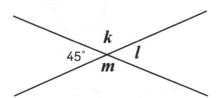

$k =$ ⬚ ° $l =$ ⬚ ° $m =$ ⬚ °

f

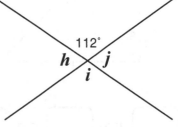

$h =$ ⬚ ° $i =$ ⬚ ° $j =$ ⬚ °

3 Calculate the size of the unknown angles.

a

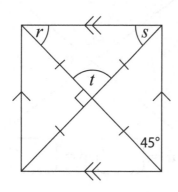

$r =$ ☐ ° $s =$ ☐ ° $t =$ ☐ °

d

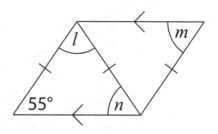

$l =$ ☐ ° $m =$ ☐ ° $n =$ ☐ °

b

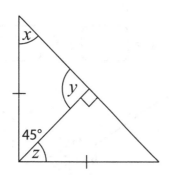

$x =$ ☐ ° $y =$ ☐ ° $z =$ ☐ °

e

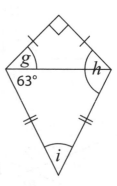

$g =$ ☐ ° $h =$ ☐ ° $i =$ ☐ °

c

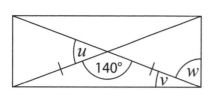

$u =$ ☐ ° $v =$ ☐ ° $w =$ ☐ °

f

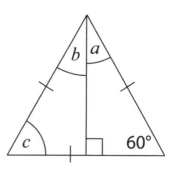

$a =$ ☐ ° $b =$ ☐ ° $c =$ ☐ °

4

YOU WILL NEED:
• ruler

Draw a quadrilateral on this grid and label the vertices and angles.

Measure or calculate the interior and exterior angles for each.

Work out other facts about your shape to answer the questions.

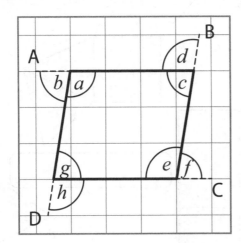

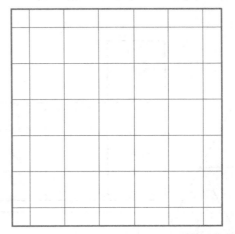

Name of your quadrilateral:

Angles

Interior angle *a*:

Exterior angle *b*:

A ➜

Interior angle *c*:

Exterior angle *d*:

B ➜

Interior angle *e*:

C ➜

Exterior angle *f*:

D ➜

Interior angle *g*:

Exterior angle *h*:

perimeter ➜ [   ] cm

area ➜ [   ] cm²

Interior angle *i*:

Exterior angle *j*:

lines of symmetry? ➜

53

1 Predict the closed shapes for each of these nets.

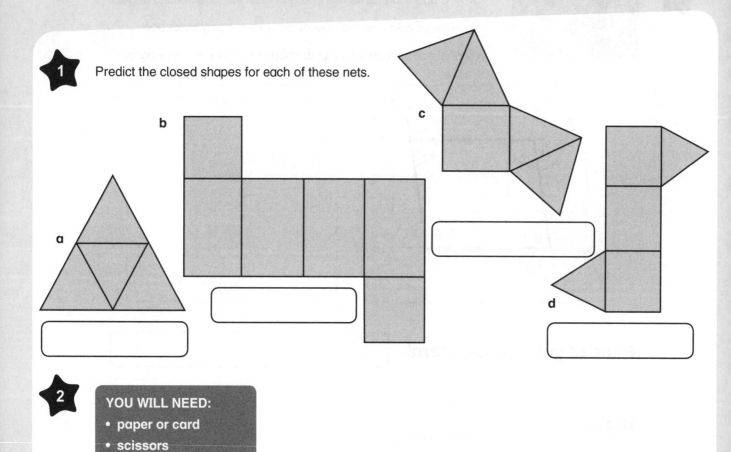

a

b

c

d

2

YOU WILL NEED:
- paper or card
- scissors

Check your predictions for question 1 by making each shape.

Open your shapes out and draw any other nets that can be made for each shape.

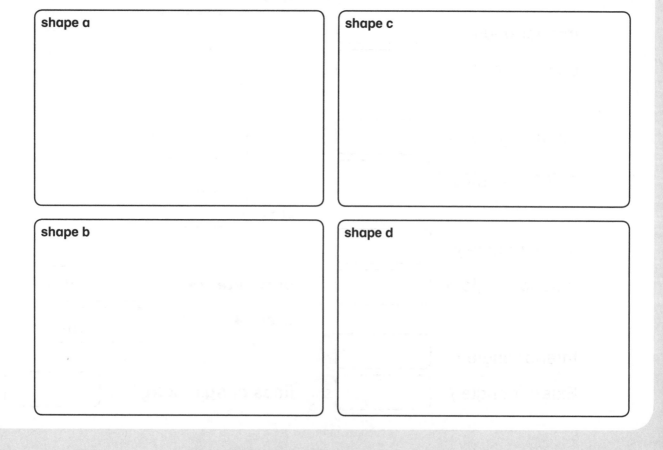

shape a

shape c

shape b

shape d

3   Calculate the volume of these cuboids.

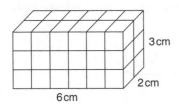

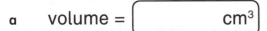

a   volume = [   ] cm³

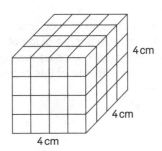

b   volume = [   ] cm³

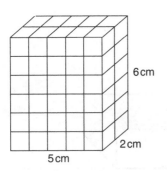

c   volume = [   ] cm³

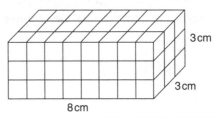

d   volume = [   ] cm³

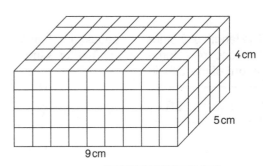

e   volume = [   ] cm³

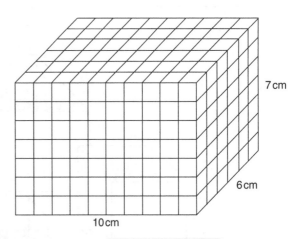

f   volume = [   ] cm³

4 Answer these.

a The area of the base of a box is 32 cm². The height of the box is 9 cm. What is the **volume** of the box?

cm³

b A cube is 7 cm high. What is the **volume** of this cube?

cm³

c The length of 2 sides of a cuboid are 3 cm and 4 cm and the volume is 60 cm³. What is the **height** of this cuboid?

cm

d The length of a box is 8 cm and the end of the box is a square with a height of 4 cm. What is the **volume** of this box?

cm³

e The volume of a box is 270 cm³ and the area of the base is 90 cm². What is the **height** of this box?

cm

f A cube has a volume of 216 cm³. What is the length of each side?

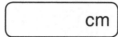
cm

g A cuboid has a square end with an area of 25 cm². The other four rectangle sides of the cuboid have sides that are twice as long as they are high. What is the **volume** of this cuboid?

cm³

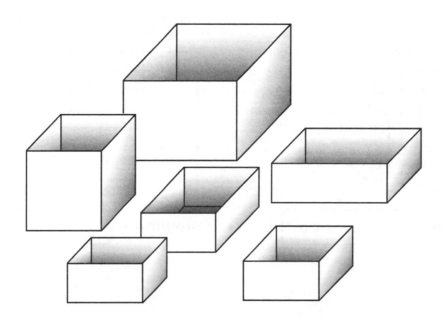

5

Design a cuboid container to hold 3 tennis balls so that they cannot move around in the box.

a Draw a sketch of your container here.

b Draw a net of your cuboid here, showing the lengths of each side.

c Make your container from card or a construction kit. Test it with 3 tennis balls.

What is the volume of your container? ☐ cm^3

Unit 5 Numbers in everyday life

5a Negative numbers in real life

1 Write the missing numbers in each sequence.

a | −11 | −7 | | 1 | | 9 | |

b | 27 | 18 | | | −9 | | −27 |

c | | −14 | −8 | | 4 | | 16 |

d | −6 | | −2 | 0 | | | 6 |

e | | 3 | −2 | | | −17 | −22 |

f | | −5 | | 3 | 7 | | 15 |

g | 10 | | 4 | | −2 | | −8 |

h | 13 | | 3 | | −7 | | −17 |

2 Show the jumps on each number line to answer these.

a 7 add −5 is [ ]

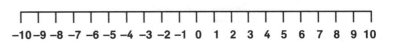

b 2 subtract 8 is [ ]

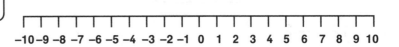

c −1 add 6 is [ ]

d 0 subtract 5 is [ ]

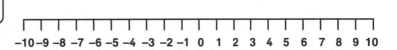

e 4 add − 4 is [ ]

f −3 subtract 7 is [ ]

g −4 add −2 is [ ]

−10 −9 −8 −7 −6 −5 −4 −3 −2 −1 0 1 2 3 4 5 6 7 8 9 10

h 1 subtract −6 is [ ]

−10 −9 −8 −7 −6 −5 −4 −3 −2 −1 0 1 2 3 4 5 6 7 8 9 10

3 Answer these.

a −5 + 6 = [ ]

b 4 − 7 = [ ]

c 0 − 9 = [ ]

d 12 − − 2 = [ ]

e −25 + 8 = [ ]

f 19 − − 11 = [ ]

g −26 + 26 = [ ]

h 33 − − 15 = [ ]

4 Complete these.

a Plot these coordinates with crosses on the grid.

(5,3) (6,−3) (0,−1) (−6,−2) (−5,4) (0,2)

b Join the coordinates in order. What shape have you drawn?

[ ]

c Plot these coordinates with crosses on the grid.

(4,−2) (−4,−2) (−4,2) (4,2)

d Which is the odd one out? Why? _____

_____

1 Convert these centimetres and metres.

a
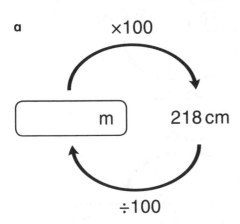

×100

m 218 cm

÷100

d
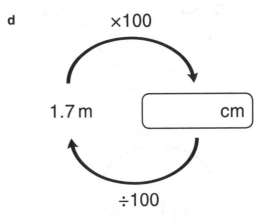

×100

1.7 m cm

÷100

b
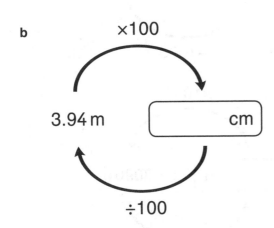

×100

3.94 m cm

÷100

e
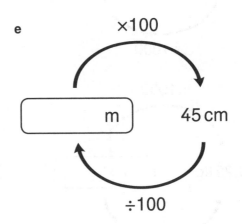

×100

m 45 cm

÷100

c
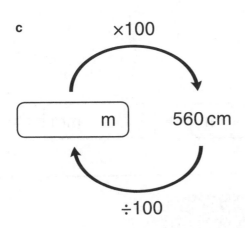

×100

m 560 cm

÷100

f
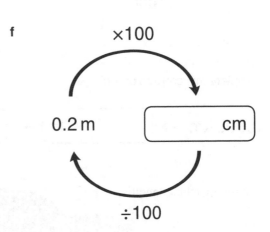

×100

0.2 m cm

÷100

2 Convert these grams and kilograms.

a
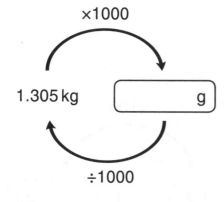
×1000

1.305 kg → [___ g]

÷1000

d
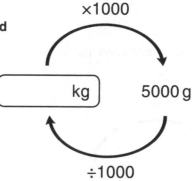
×1000

[___ kg] ← 5000 g

÷1000

b
×1000

[___ kg] ← 4018 g

÷1000

e
×1000

8.6 kg → [___ g]

÷1000

c
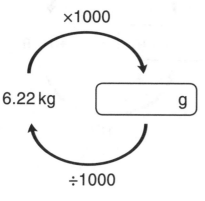
×1000

6.22 kg → [___ g]

÷1000

f
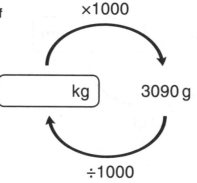
×1000

[___ kg] ← 3090 g

÷1000

 3

a Complete this conversion trail.

0.008 km → [___ m] → [___ cm] → [___ mm]

b Join the matching lengths.

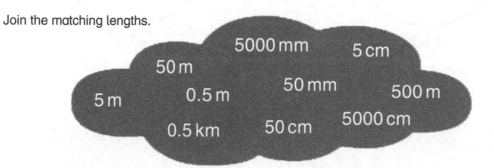

5000 mm   5 cm

50 m

5 m   0.5 m   50 mm   500 m

0.5 km   50 cm   5000 cm

4

a These are some of the top 1500 m times for British female athletes. Convert the times to seconds and complete this chart.

First and last name	Personal best time (minutes)	Personal best time (seconds)
Jessica Judd	4:09.56	
Rachael Bamford	4:12.59	
Madeleine Murray	4:10.17	
Melissa Courtney	4:09.74	
Rosie Clarke	4:12.10	
Laura Weightman	4:00.17	
Kelly Holmes	3:57.90	
Laura Muir	3:58.66	

b Write the runners in order, starting with the **fastest**. Write the last name and the time in seconds for each athlete.

Last name	Personal best time (seconds)
Holmes	
Muir	
Weightman	
Judd	
Courtney	
Murray	
Clarke	
Bamford	

5 Water is leaking from a tap at a rate of 0.07 litres every 5 seconds.

a How much water is lost in 1 minute?

| ml |

c How much water is lost in 1 day?

| litres |

b How much water is lost in 1 hour?

| litres |

d How long will it take to fill a 1 litre jug? Give your answer to 2 decimal places.

| seconds |

0.07 litres × ★ = 1 litre → ★ × 5 seconds =

Solving problems

1 Draw the jumps on each number line back to zero to help answer each of these.

$23 - 39 =$ $\boxed{-16}$

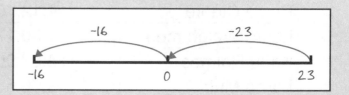

a $15 - 32 =$ ⬚

0   15

b $27 - 40 =$ ⬚

0   27

c $14 - 54 =$ ⬚

0   14

d $33 - 76 =$ ⬚

0   33

e $29 - 65 =$ ⬚

0   29

f $47 - 72 =$ ⬚

0   47

2 This table shows the temperatures during the day and at night for a week up a mountain. Complete the table to show the difference between each day and night temperature. Use the bead string to help calculate the differences.

	Day temperature	Night temperature	Difference
Monday	7 °C	−12 °C	
Tuesday	11 °C	−9 °C	
Wednesday	15 °C	−11 °C	
Thursday	9 °C	−7 °C	
Friday	8 °C	−8 °C	
Saturday	5 °C	−12 °C	
Sunday	6 °C	−13 °C	

3 This bar graph shows the hottest and coldest temperatures reached in a year in these cities. The top of the bar shows the hottest temperature and the bottom of each bar shows the coldest temperature. Use the graph to answer the questions.

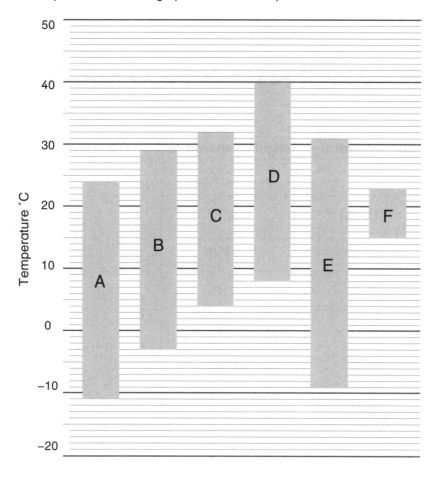

A Moscow

B New York

C Rome

D New Delhi

E Beijing

F London

a Which city reached the hottest temperature?

b What was the coldest temperature reached in Rome?

 °C

c Which city had the largest temperature change?

d Which city had the smallest temperature change?

e What was the difference between the hottest and coldest temperature in Rome?

 °C

f What was the difference between the hottest and coldest temperature in Moscow?

 °C

g Which city had a temperature change of 18°C?

h Which 2 cities had the same temperature change?

and

4 These are magic squares.

a Each column, row and diagonal of this magic square adds to −12. Write the missing number.

−2	0	−10
−12		4
2	−8	−6

c Write the missing numbers. What does each column, row and diagonal add to?

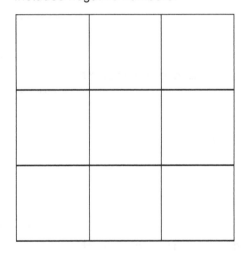

		3
2		−2
		−1

b Write the missing numbers. What does each column, row and diagonal add to?

−11		−7
	−5	
−3		

d Make up your own magic square which includes negative numbers.

1 What change from £20 would you receive if you bought each pair of items?

a Change from £20 ➜ [   ]

£9.50 each

b Change from £20 ➜ [   ]

£3.80 each

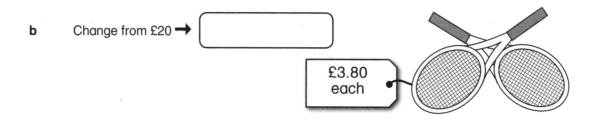

c Change from £20 ➜ [   ]

£7.40 each

d Change from £20 ➜ [   ]

£4.99 each

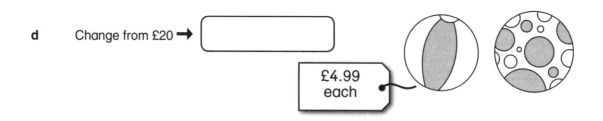

e Change from £20 ➜ [   ]

£8.85 each

2 Calculate the cost of each set. Which set costs more and by how much?

a Which costs more, 4 cakes or 6 biscuits?

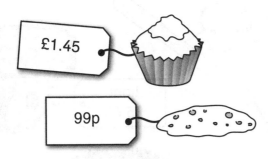

£1.45

99p

costs more by £ [   ]

b Which costs more, 5 cones or 6 lollies?

£1.29

£1.19

costs more by £ [   ]

c Which costs more, 4 pizza slices or 5 rolls?

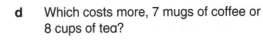

£2.96

£2.29

costs more by £ [   ]

d Which costs more, 7 mugs of coffee or 8 cups of tea?

£1.89

£2.09

costs more by £ [   ]

These are the prices for some items in a bicycle shop.

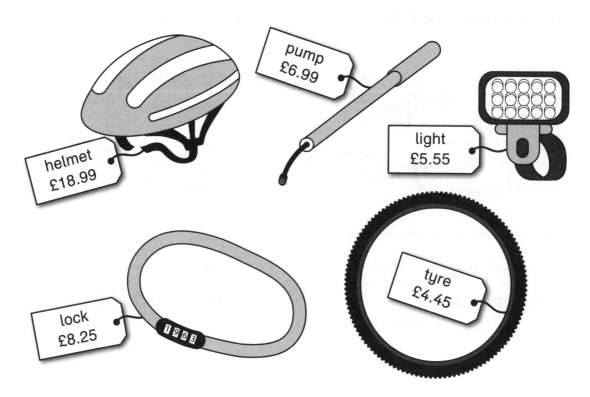

helmet
£18.99

pump
£6.99

light
£5.55

lock
£8.25

tyre
£4.45

a What is the cost of a pump and 2 tyres?

£

b What is the cost of a pair of lights and a lock?

£

c What is the cost of a helmet, a pump and a lock?

£

d What change would you get from £20
if you bought a pump and a tyre?

£

e Which costs more, 2 tyres and a lock or
2 lights and a pump?
How much more do they cost?

£

4 When 4 friends went shopping, each had £30 to spend. At a card shop they each bought some cards. Write how much they each spent on cards and how much money each had left from £30.

These codes show the prices for each card.

Code	A	B	C	D	E
Price	£2.05	£1.63	£3.42	£1.30	£2.18

a

Total cost £

Change from £30 £

c

Total cost £

Change from £30 £

b

Total cost £

Change from £30 £

d

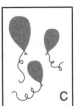 

Total cost £

Change from £30 £

1 Write the place value of **each** digit in these numbers.

$$4.86 = 4 + \frac{8}{10} + \frac{5}{100} + \frac{6}{1000}$$

a 2.476 =

d 12.35 =

b 9.315 =

e 14.907 =

c 0.048 =

f 29.937 =

2 Round each of the numbers to the nearest **tenth**.

a 2.476 =

d 12.35 =

b 9.315 =

e 14.907 =

c 0.058 =

f 29.937 =

3 Round each of these to the nearest **whole** number.

a 18.48 cm

b 25.389 kg

c 10.71 m

d 32.503 litres

e 29.08 km

f 17.6 g

g 0.94 m

h 46.199 km

i 50.273 kg

j 26.81 cm

4 Change these fractions to decimals.

$$\frac{7}{8} = 8\overline{)7.000} \quad 0.875$$

a $\frac{4}{5}$

b $\frac{1}{8}$

c $\frac{3}{5}$

d $\frac{3}{4}$

5 Some fractions do not change exactly into decimals.
Change these to decimals rounded to **3** decimal places.

a $\frac{1}{6}$

c $\frac{7}{9}$

b $\frac{3}{11}$

d $\frac{4}{7}$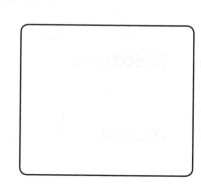

 6

YOU WILL NEED:
- **3 dice**
- **partner**

Play a game of 'Round and Win Bingo' with a partner. You need 3 dice.
- First each player writes 6 different whole numbers between 1 and 7 in their grid below.

Player 1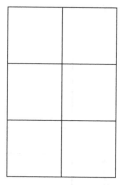

Player 2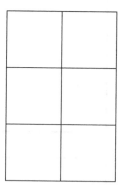

- Then take turns to roll the 3 dice all at once, and arrange them in these boxes so they make a number with 2 decimal places.

- Round the number you have made to the nearest whole number. If that number is on your grid, cross it off.
- The first player to cross out a column scores 1 point and the first to cross out the whole grid scores 3 points.
- Play the game 3 times to find the winner – best out of 3 wins!

1 Here are some rules for sequences of numbers.
Write the first 6 numbers in each sequence.

a Start at 2. Add 3 each time.

| 2 | 5 | | | | |

b Start at 18. Subtract 5 each time.

| 18 | | | | | |

c Start at −5. Add 6 each time.

| −5 | | | | | |

d Start at 1. Add 9 each time.

| 1 | | | | | |

Use rods to make this sequence. Continue the sequence up to the 6th term.

1st term

2nd term

3rd term

4th term

5th term

6th term

a   Complete this table

Term number (n)	1	2	3	4	5	6
Rods (r)	3					

b   What patterns do you notice?

3 A formula for the rods pattern could be $r = 2n + 1$. Use this to work out the number of rods for these terms:

a   10th term ➡️ ⬜

b   15th term ➡️ ⬜

c   30th term ➡️ ⬜

d   100th term ➡️ ⬜

e   18th term ➡️ ⬜

f   54th term ➡️ ⬜

g   Which term in the sequence will have a total of 51 rods?

th term

4

YOU WILL NEED:
• rods (optional)

Use rods to make the next 3 terms in each of these sequences. Record the results and work out a formula for each.

a

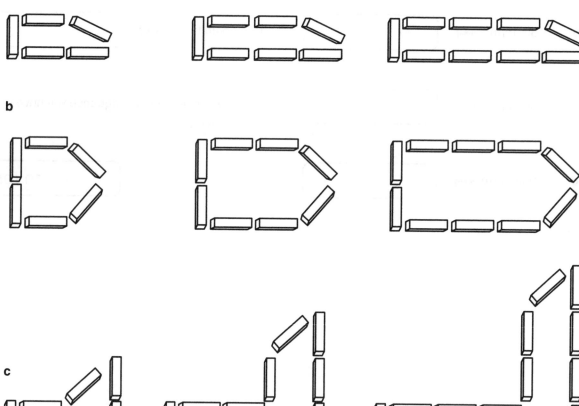

b

c

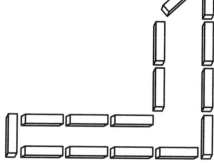

a

Term number (n)	1	2	3	4	5	6
Rods (r)						

formula: [   ]

b

Term number (n)	1	2	3	4	5	6
Rods (r)						

formula: [   ]

c

Term number (n)	1	2	3	4	5	6
Rods (r)						

formula: [   ]

a   Make a linear sequence that has this formula.

To do this, use counters or coins and continue this pattern.

Term number (n)	1	2	3	4
⬤ (small)	3	4	5	6
⬤ (large)	2	3	4	5
Total	5	7	9	11

formula: total = $2n + 3$

b   Make up your own linear sequence with counters.

Complete the table to show your formula for the pattern.

Term number (n)	1	2	3	4
⬤ (small)				
⬤ (large)				
Total				

formula: total =

Let's explore fractions and algebra!

7a Fraction equivalences

 1 Complete these fraction equivalent families.

a $\dfrac{2}{3}$ $\dfrac{4}{\Box}$ $\dfrac{6}{\Box}$ $\dfrac{\Box}{12}$ $\dfrac{10}{\Box}$ $\dfrac{\Box}{18}$

b $\dfrac{3}{4}$ $\dfrac{\Box}{8}$ $\dfrac{\Box}{12}$ $\dfrac{12}{\Box}$ $\dfrac{\Box}{20}$ $\dfrac{18}{\Box}$

c $\dfrac{2}{5}$ $\dfrac{4}{\Box}$ $\dfrac{\Box}{15}$ $\dfrac{\Box}{20}$ $\dfrac{10}{\Box}$ $\dfrac{\Box}{30}$

d $\dfrac{4}{5}$ $\dfrac{\Box}{10}$ $\dfrac{12}{\Box}$ $\dfrac{\Box}{20}$ $\dfrac{20}{\Box}$ $\dfrac{\Box}{30}$

 2 Shade each grid to show the equivalent fractions. Draw a circle round the largest fraction.

a $\dfrac{2}{3}$ 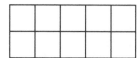 $\dfrac{5}{6}$

b $\dfrac{2}{5}$ $\dfrac{3}{10}$

c $\frac{1}{2}$ 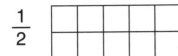 $\frac{3}{5}$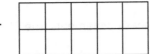

d $\frac{1}{3}$ $\qquad$ $\frac{1}{4}$

e $\frac{5}{6}$ $\qquad$ $\frac{3}{4}$

3 Calculate the equivalent fractions for each pair. Then write < or > to compare the fractions.

a

$\times \boxed{}$

$\frac{2}{3} = \dfrac{\boxed{}}{\boxed{}}$

$\times \boxed{}$

$\times \boxed{}$

$\frac{3}{5} = \dfrac{\boxed{}}{\boxed{}}$

$\times \boxed{}$

$\frac{2}{3} \; \boxed{} \; \frac{3}{5}$

b

$\times \boxed{}$

$\frac{3}{4} = \dfrac{\boxed{}}{\boxed{}}$

$\times \boxed{}$

$\times \boxed{}$

$\frac{5}{6} = \dfrac{\boxed{}}{\boxed{}}$

$\times \boxed{}$

$\frac{3}{4} \; \boxed{} \; \frac{5}{6}$

c

$\times \boxed{}$

$\frac{5}{8} = \dfrac{\boxed{}}{\boxed{}}$

$\times \boxed{}$

$\times \boxed{}$

$\frac{2}{3} = \dfrac{\boxed{}}{\boxed{}}$

$\times \boxed{}$

$\frac{5}{8} \; \boxed{} \; \frac{2}{3}$

d

$\times \boxed{}$

$\frac{7}{9} = \dfrac{\boxed{}}{\boxed{}}$

$\times \boxed{}$

$\times \boxed{}$

$\frac{5}{6} = \dfrac{\boxed{}}{\boxed{}}$

$\times \boxed{}$

$\frac{7}{9} \; \boxed{} \; \frac{5}{6}$

4 Write <, > or = to make these statements true.

a $1\frac{3}{5}$ ☐ $1\frac{2}{3}$

d $5\frac{2}{5}$ ☐ $5\frac{3}{10}$

b $3\frac{3}{4}$ ☐ $\frac{15}{4}$

e $\frac{35}{8}$ ☐ $4\frac{3}{8}$

c $3\frac{1}{2}$ ☐ $\frac{31}{10}$

f $\frac{6}{5}$ ☐ $\frac{7}{6}$

 5 Write these lengths in order, starting with the **shortest**.

a $\frac{3}{4}$ km $\frac{2}{3}$ km $\frac{1}{2}$ km $\frac{5}{6}$ km

☐ ☐ ☐ ☐
Shortest

b $\frac{5}{8}$ m $\frac{1}{2}$ m $\frac{3}{8}$ m $\frac{3}{4}$ m

☐ ☐ ☐ ☐
Shortest

c $\frac{5}{6}$ km $\frac{3}{4}$ km $\frac{2}{3}$ km $\frac{7}{8}$ km

☐ ☐ ☐ ☐
Shortest

d $1\frac{1}{2}$ km $2\frac{3}{10}$ km $2\frac{1}{4}$ km $1\frac{2}{5}$ km

☐ ☐ ☐ ☐
Shortest

e $3\frac{3}{4}$ cm $2\frac{5}{8}$ cm $3\frac{5}{6}$ cm $2\frac{2}{3}$ cm

☐ ☐ ☐ ☐
Shortest

f $2\frac{3}{8}$ m $3\frac{2}{3}$ m $3\frac{3}{4}$ m $2\frac{5}{6}$ m

☐ ☐ ☐ ☐
Shortest

6

Add and subtract these fractions. Simplify the answers if possible. Change them to equivalent fractions with a common denominator. Use the strips of boxes to help you.

a $\dfrac{2}{3} + \dfrac{3}{5} =$

b $\dfrac{5}{6} + \dfrac{3}{4} =$

c $\dfrac{3}{4} + \dfrac{2}{5} =$

d $\dfrac{4}{9} + \dfrac{1}{2} =$

e $\dfrac{4}{5} - \dfrac{1}{2} =$

f $\dfrac{3}{4} - \dfrac{1}{6} =$

g $\dfrac{2}{3} - \dfrac{2}{5} =$

h $\dfrac{1}{2} - \dfrac{4}{9} =$

These 3 cases were weighed before going on a flight.

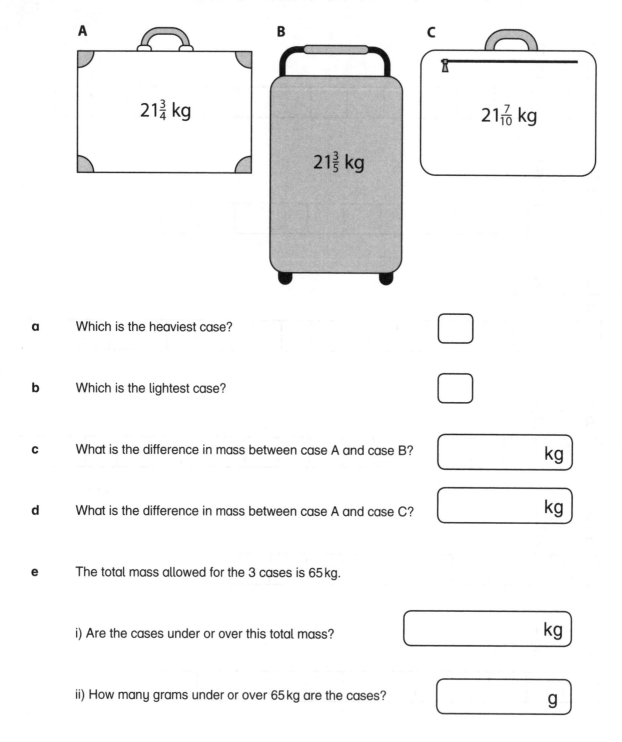

A $21\frac{3}{4}$ kg

B $21\frac{3}{5}$ kg

C $21\frac{7}{10}$ kg

a Which is the heaviest case?

b Which is the lightest case?

c What is the difference in mass between case A and case B? kg

d What is the difference in mass between case A and case C? kg

e The total mass allowed for the 3 cases is 65 kg.

i) Are the cases under or over this total mass? kg

ii) How many grams under or over 65 kg are the cases? g

Fraction, decimal and percentage equivalences

 1 Change these 6 test scores in each set to percentages.

a Test out of 10

$\frac{9}{10}$	$\frac{7}{10}$	$\frac{6}{10}$	$\frac{5}{10}$	$\frac{3}{10}$	$\frac{1}{10}$
90 %					

b Test out of 20

$\frac{18}{20}$	$\frac{14}{20}$	$\frac{12}{20}$	$\frac{10}{20}$	$\frac{6}{20}$	$\frac{2}{20}$

c Test out of 20

$\frac{19}{20}$	$\frac{15}{20}$	$\frac{11}{20}$	$\frac{5}{20}$	$\frac{12}{20}$	$\frac{4}{20}$

 2 Write the missing fractions, decimals and percentages in this chart.

$\frac{1}{2}$	$\frac{1}{5}$				$\frac{2}{5}$
50 %		75 %		60 %	
0.5			0.02		

3 Rewrite these newspaper headlines using **percentages**.

a Eight out of ten cats prefer a quiet night in watching the TV.

_____

_____

b There was a one-in-two chance that the goldfish would swim to the left.

_____

_____

c Out of the fifty people questioned, thirty-two of them said they didn't do it.

_____

_____

d Only seven out of the twenty-five diners thought the food was good enough to eat.

_____

_____

e Four in five people read a daily paper, with three in ten of those just reading the sport.

_____

_____

4 Use the first percentage to help work out the others.

a **10% of £12 =** ☐

20% of £12 = ☐

30% of £12 = ☐

70% of £12 = ☐

5% of £12 = ☐

b **1% of £12 =** ☐

2% of £12 = ☐

3% of £12 = ☐

7% of £12 = ☐

5% of £12 = ☐

c **50% of £12 =** ☐

25% of £12 = ☐

5% of £12 = ☐

55% of £12 = ☐

30% of £12 = ☐

5 Answer these.

a A company makes 2500 car parts each week.

30% of the parts are exported.

How many car parts are exported?

b A designer ordered fabric costing £1250.

She paid a deposit of 20%.

How much deposit did she pay?

£

c A 2 litre bottle of juice is made with 40% fruit and 60% water.

How many millilitres of water is there in this bottle?

ml

d In a cinema during half-term week only 25% of

the 5400 audience were adults.

The rest were children.

How many adults went to the cinema during half-term?

e Grandma saved £1900 to take her grandchildren on holiday.

She kept the money for a year in the bank where

it earned 5% interest. How much money did she get in

interest by the end of the year?

£

1 The perimeter of a rectangle can be calculated by adding the length and width and then doubling the total. The formula for this is $p = 2(l + w)$

Use the formula to work out the perimeter of these rectangles.

a

3.5 cm

8 cm

perimeter = ☐ cm

b

70 mm

40 mm

perimeter = ☐ mm

c

12.5 m

4.5 m

perimeter = ☐ m

d

800 mm

600 mm

perimeter = ☐ mm

e

9.25 cm

4.25 cm

[cm]

f

550 mm

350 mm

[mm]

2 Answer these.

a The perimeter of a rectangle is 13 cm and the length of the longest side is 4.2 cm. What is the width of the rectangle?

[cm]

b The perimeter of a rectangle is 25 cm and the width is 4 cm. What is the length of the rectangle?

[cm]

c The length of a rectangle is double the width. If the perimeter of the rectangle is 18 cm, what is the length and width?

[cm]

and

[cm]

d The width of a rectangle is 25% of the length. If the perimeter of the rectangle is 50 cm, what is the length and width?

[cm]

and

[cm]

89 ★

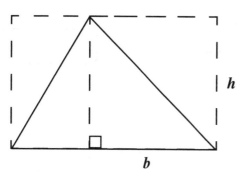 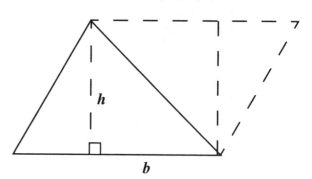

We use formulae to find the area of triangles and parallelograms.

a is area, b is base length, h is height.

The formula for the area of a triangle is $a = \frac{bh}{2}$

The formula for the area of a parallelogram is $a = bh$

Calculate the area of these triangles and parallelograms.

a

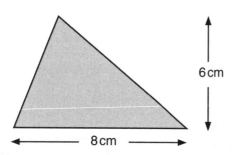

6 cm

8 cm

area of triangle = ☐ cm²

b

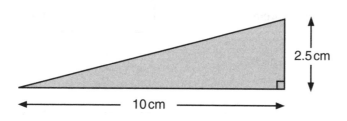

2.5 cm

10 cm

area of triangle = ☐ cm²

c

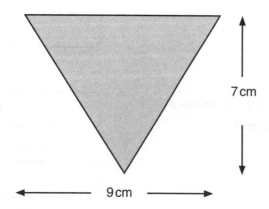

7 cm

9 cm

area of triangle = ☐ cm²

d

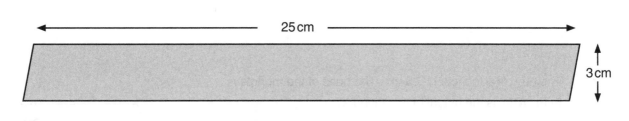

5 cm

9 cm

area of parallelogram = [　　　] cm^2

e

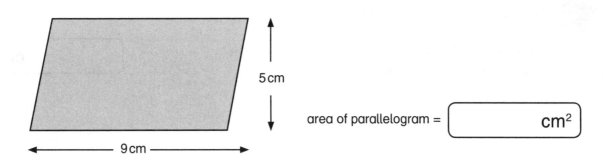

25 cm

3 cm

area of parallelogram = [　　　] cm^2

f

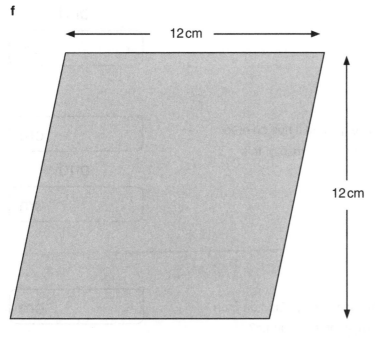

12 cm

12 cm

area of parallelogram = [　　　] cm^2

4 Use the formulae to calculate these.

a The area of a parallelogram is 36 cm². Its height is 4 cm. What is the length of the base?

[ ] cm

b The perimeter of a square is 32 cm. What is the area of this square?

[ ] cm²

c The area of a triangle is 7.5 cm². The base of the triangle is 5 cm. What is the height of this triangle?

[ ] cm

d The area of a rectangle is 24 cm² and the perimeter is 22 cm. What is the length and width of this rectangle?

[ ] cm

and

[ ] cm

e The base of a triangle is double its height and it has an area of 9 cm². What is the height and length of the base of this triangle?

[ ] cm

and

[ ] cm

f A triangle has a height and base the same length and an area of 18 cm². What is the height and length of this triangle?

[ ] cm

and

[ ] cm

5   A square is joined to a rectangle to make an L-shape.

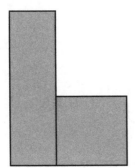

The area of the L-shape is 36 cm². Each side is a whole number of centimetres.

Use this chart to help you investigate the shortest and longest perimeter of the L-shape when the square has the areas below:

Area of square (sides in cm)	Area of rectangle	Possible length of sides of rectangle	Shortest perimeter of L-shape	Longest perimeter of L-shape
4 cm² (sides: 2 cm)	32 cm²	1 cm × 32 cm 2 cm × 16 cm 4 cm × 8 cm		
9 cm² (sides: 3 cm)	27 cm²			
16 cm² (sides: ☐ cm)	20 cm²			
25 cm² (sides: ☐ cm)	11 cm²			

1   Write the missing numbers.

a $17 +$ ☐ $= 31$

b $14 +$ ☐ $= 31$

c   ☐ $+ 12 = 48$

d $24 +$ ☐ $= 48$

e   ☐ $+ 36 = 64$

f $56 +$ ☐ $= 84$

g   ☐ $+ 49 = 91$

h $29 +$ ☐ $= 91$

2   Use the bar model to help find the unknown number.

a		b
c	d	

$a + 8 = 10 + 15$ 10 add 15 is 25.

$a = \boxed{17}$ Now take away 8 from 25 and a is 17.

a $17 + b = 9 + 16$

   $b =$ ☐

b $a + 15 = 18 + 31$

   $a =$ ☐

c $26 + 15 = c + 30$

   $c =$ ☐

d $43 + 27 = 23 + d$

   $d =$ ☐

3 Write the numbers from question 2 on to these weights. Check the equation is balanced.

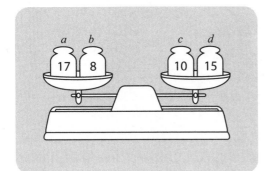

a

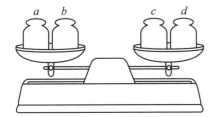

c

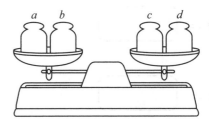

b

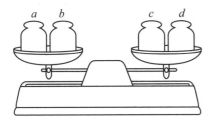

d

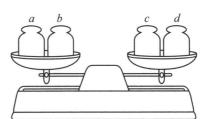

4 Calculate the value of the unknown number.

a $n + 15 = 31$ $n =$ ⬚   **e** $54 + 12 = 100 - n$ $n =$ ⬚

b $18 - n = 9 + 5$ $n =$ ⬚   **f** $n - 9 = 7 \times 3$ $n =$ ⬚

c $3n - 6 = 18$ $n =$ ⬚   **g** $22 + n = 7^2$ $n =$ ⬚

d $36 \div 4 = n - 7$ $n =$ ⬚   **h** $18 + 3n = 6 \times 4$ $n =$ ⬚

5

Here are three equations.

$$a + b + c = 48$$
$$a + b = 27$$
$$b + c = 36$$

What are the values of a, b and c?

$a =$ [ ]

$b =$ [ ]

$c =$ [ ]

8a Identifying common factors, multiples and prime numbers

1 Read each arrow statement. Draw arrows to join sets of numbers. Some numbers will be joined to more than one number or none at all.

a is half of

$\longrightarrow$

9	12	18
11	24	22
6	36	

c ×2 then +1

$\longrightarrow$

5	11	3
7	15	23
9	18	

b is a factor of

$\longrightarrow$

3	4	2
9	18	7
16	20	

d is a multiple of

$\longrightarrow$

15	3	30
14	7	8
24	56	

Check that the direction of your arrow is correct.

2 The direction of each arrow in question 1 is reversed. What would you write on the arrow now?

a

$\longleftarrow$

c

$\longleftarrow$

b

$\longleftarrow$

d

$\longleftarrow$

3

a Write these numbers in the correct parts of this Venn diagram.

| 3 20 12 66 54 18 72 81 48 36 |

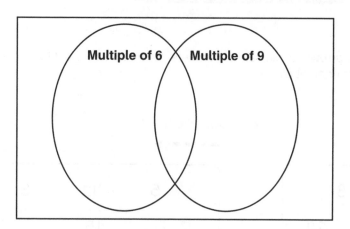

Multiple of 6   Multiple of 9

b Which numbers in this set are the common multiples of 6 and 9?

c Which numbers in this set are **not** multiples of either 6 or 9?

4

a Write these numbers in the correct part of this Carroll diagram.

| 1 2 3 4 5 6 8 10 12 15 16 24 32 |

	Factor of 32	Not factor of 32
factor of 24		
not factor of 24		

b Does this show all the factors of 24 and also all the factors of 32? Answer yes or no.

d Which numbers in this set are the common factors of 24 and 32?

c Which numbers in this set are not factors of 24 or of 32?

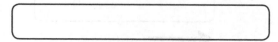
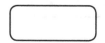

5 Simplify these fractions by finding common factors.

a $\dfrac{20}{35} = \boxed{}$

d $\dfrac{18}{72} = \boxed{}$

g $\dfrac{20}{96} = \boxed{}$

b $\dfrac{14}{21} = \boxed{}$

e $\dfrac{15}{95} = \boxed{}$

h $\dfrac{26}{70} = \boxed{}$

c $\dfrac{9}{42} = \boxed{}$

f $\dfrac{21}{90} = \boxed{}$

i $\dfrac{45}{60} = \boxed{}$

6 Here is a factor tree for 60.
60 is the product of three prime numbers: 2, 3 and 5.
These are the prime factors of 60.

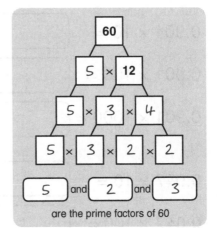

Work out the prime factors of these numbers. Draw factor trees like the example above.

a 42

$\boxed{}$ and $\boxed{}$ and $\boxed{}$

are the prime factors of 42

c 90

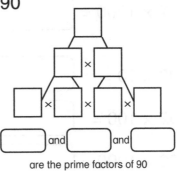

$\boxed{}$ and $\boxed{}$ and $\boxed{}$

are the prime factors of 90

b 72

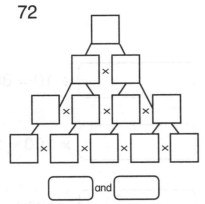

$\boxed{}$ and $\boxed{}$

are the prime factors of 72

d 132

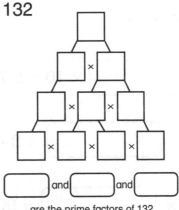

$\boxed{}$ and $\boxed{}$ and $\boxed{}$

are the prime factors of 132

1 Answer these.

a 0.385 × 10 = [ ]

0.385 × 100 = [ ]

0.385 × 1000 = [ ]

b 0.901 × 10 = [ ]

0.901 × 100 = [ ]

0.901 × 1000 = [ ]

c 0.047 × 10 = [ ]

0.047 × 100 = [ ]

0.047 × 1000 = [ ]

d 0.002 × 10 = [ ]

0.002 × 100 = [ ]

0.002 × 1000 = [ ]

e 219 ÷ 10 = [ ]

219 ÷ 100 = [ ]

219 ÷ 1000 = [ ]

f 63 ÷ 10 = [ ]

63 ÷ 100 = [ ]

63 ÷ 1000 = [ ]

g 50 ÷ 10 = [ ]

50 ÷ 100 = [ ]

50 ÷ 1000 = [ ]

h 8 ÷ 10 = [ ]

8 ÷ 100 = [ ]

8 ÷ 1000 = [ ]

2 Complete these.

a [ ] × 10 = 5.7

b 49 ÷ [ ] = 0.49

c 0.9 × [ ] = 900

d [ ] ÷ 10 = 30.3

e [ ] × 100 = 3.6

f [ ] ÷ 1000 = 0.028

3 Answer these.

a   3 . 6 7 2
    ×   4

b   1 . 4 8 5
    ×   6

c   2 . 0 3 7
    ×   5

d   1 0 . 9 6
    ×   7

e   1 7 . 5 4
    ×   8

f   2 1 . 9 0 3
    ×   3

4 Answer these.

a   3⟌6.426

b   4⟌7.428

c   3⟌9.255

d   5⟌8.169

e   6⟌4.326

f   8⟌5.104

5 Calculate the area of these rectangles.

a

7.25 cm

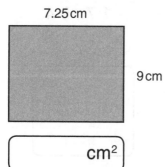

9 cm

cm²

b

6 cm

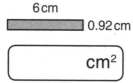0.92 cm

cm²

c

12 cm

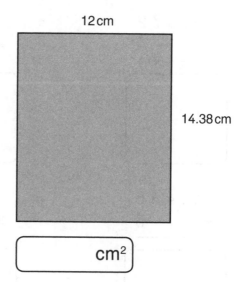

14.38 cm

cm²

d

11 cm

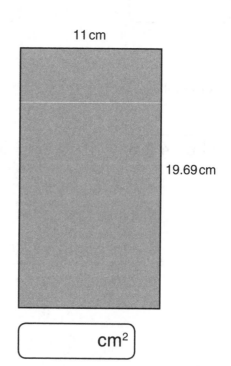

19.69 cm

cm²

6 Answer these.

a   The distance between Eva's home and her school is 1.235 km. She walks to school each morning and back home each evening. She does this every day from Monday to Friday. How far does she walk in total in 1 week?

[ ] km

b   A ticket to a theme park costs £24.79. How much will it cost in total for 3 people to visit the theme park?

£ [ ]

c   A length of pipe is 3.74 m. A trench is dug alongside a road and 40 lengths of pipe are placed end to end in the trench. What is the total length of the pipes ?

[ ] m

d   1.74 litres of juice is poured equally into 6 glasses. How much juice is in each glass?

[ ] litres

e   A pack of 4 red T-shirts cost £9.48 and a pack of 5 blue T-shirts costs £11.75. Which costs less, 1 red T-shirt or 1 blue T-shirt?

[ ]

f   Ali makes some cakes. The recipe is shown below. Ali shares the mixture equally between 12 holes on a baking tray. What is the mass of each cake?

[ ] kg

Cake recipe:

1 Kg flour

0.645 Kg butter

0.077 Kg nuts

0.21 Kg eggs

a Complete this grid.

	10%	5%	20%	2%
£30	£3			
£12				
£46				
£80				
£94				

b Describe and show how you worked out 2% of £94.

 Answer these.

a 30% of 250 = ☐   **d** 25% of 360 = ☐

b 70% of 310 = ☐   **e** 11% of 420 = ☐

c 90% of 150 = ☐   **f** 3% of 210 = ☐

3 Calculate the reduced price of each sale item.

a

reduced by **20%**

£12.50

New price £ [   ]

d

£47.50

reduced by **20%**

New price £ [   ]

b

£15.00

reduced by **40%**

New price £ [   ]

e

reduced by **5%**

£65.00

New price £ [   ]

c

reduced by **30%**

£8.00

New price £ [   ]

4 Use bar models to help calculate these.

> **What percentage of 50 is 30?**
>
> 0%   60%   100%
>
> | | | | | |
> |---|---|---|---|---|
>
> 0   30   50

a What percentage of 30 is 12?

b What percentage of 45 is 27?

c What percentage of 50 is 35?

d What percentage of 60 is 9?

e What percentage of 56 is 14?

f What percentage of 68 is 17?

5 This shows the number of visitors to a farm one weekend.

Visitors	Saturday	Sunday
adult (male)	30	40
adult (female)	36	56
children	54	64
Total	**120**	**160**

a What percentage of the total number of visitors on Saturday were male adults?

c What percentage of the total number of visitors on Sunday were female adults?

b What percentage of the total number of visitors on Saturday were children?

d What percentage of the total number of visitors on Sunday were adults?

6 Use bar models to help you answer these.

A DVD is reduced in a sale by 75%. It now costs £3. What was the original price?

£3			

If 75% is taken off the whole 100% original price, that leaves 25%. So the reduced sale price of £3 is 25% of the original price. 25% is ¼ of the whole and each part is £3. This means the original cost was £12.

a A massive 70% is taken off the price of a tent (because it leaked!). Scott didn't mind getting wet and only paid £18 for it. What was the price of the tent before the reduction?

b There is a 20% discount on all hats. A top hat now costs £76. What was the original price?

c A painting was sold in an auction. 10% of the sale price was kept by the auctioneer, which left £3600 for the seller. How much did the painting sell for?

d A survey found that 80% of cats preferred fresh fish to tinned food. 30 cats chose tinned food. How many cats were in the survey altogether?

1 Work out these unknowns.

a $c + 8 = 14$

$c = \boxed{}$

d $4n + 2 = 30$

$n = \boxed{}$

b $15 - y = 9 + 4$

$y = \boxed{}$

e $19 = 3a - 5$

$a = \boxed{}$

c $17 + 6 = d - 4$

$d = \boxed{}$

f $2c - 4 = 6 \times 3$

$c = \boxed{}$

2 Complete the table to show all the possible positive whole numbers for c and d.

a $c + d = 8$

8	
c	d

c	d
1	7

b What is the value of c if $c + d = 8$ and $c - d = 4$?

$c = \boxed{}$

c What is the value of d if $c + d = 8$ and $c - d = 2$?

$d = \boxed{}$

3 Complete the chart to help you answer this problem.

Dev bought some ice-creams that cost 80p each and some lollies that cost 50p each. He spent 10p more on the lollies than on the ice-creams and got 10p change from £5. How many of each did he buy?

ice-creams = ⬚   lollies = ⬚

Number of each	Ice-creams (80p)	Lollies (50p)
1		
2		
3		
4		
5		
6		
7		
8		

4 Answer these.

a I am thinking of two numbers. One is double the other and their total is 30. What are my two numbers?

⬚ and ⬚

b I am thinking of two numbers. The difference between them is 8 and their total is 40. What are my two numbers?

⬚ and ⬚

c I am thinking of two numbers. One is a third of the other number and their total is 20. What are my two numbers?

⬚ and ⬚

d I am thinking of two numbers. When added together they make a square number less than 30 and the difference between them is a square number less than 10. What are my two numbers?

⬚ and ⬚

e I am thinking of three consecutive numbers and their total is 45. What are my three numbers?

⬚ , ⬚ and ⬚

Shapes and coordinates

1 Label the parts of this circle.

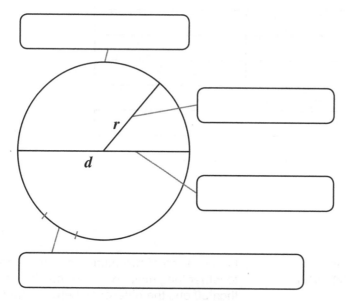

2 Complete this table to show the radius and diameter of different circles.

a

Radius (r)	Diameter (d)
8 cm	
	1.8 m
4.5 cm	
	35 cm
0.06 m	
	3.28 m

b   Draw circles around the 2 correct rules.

$d = 2r$ $r = 2d$ $d = r/2$ $r = d/2$

3

Use a pair of compasses to draw a set of 5 concentric circles in this box.

The radius of the smallest circle is 2 cm.

The radius of each circle increases by 0.5 cm.

What is the diameter of the largest circle?

4

a Calculate the area of these triangles. Complete the table to show the area of the triangles when they are increased by a scale factor of 2.

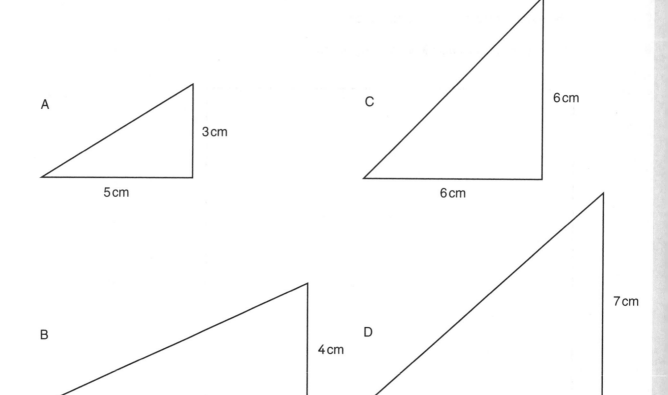

A

3 cm

5 cm

C

6 cm

6 cm

B

9 cm

4 cm

D

7 cm

8 cm

Triangle	Height (cm)	Base length (cm)	Area (cm²)	New height (cm)	New base (cm)	New area (cm²)
A	3	5				
B	4	9				
C	6	6				
D	7	8				

What do you notice?

b What is the area of each triangle if they are increased by a scale factor of 3?

A ➡ [ ] cm²

B ➡ [ ] cm²

C ➡ [ ] cm²

D ➡ [ ] cm²

1 Calculate the missing angles on these lines.

a

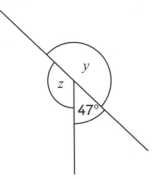

$y = \boxed{}°$ $z = \boxed{}°$

d

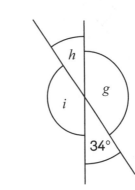

$g = \boxed{}°$ $h = \boxed{}°$ $i = \boxed{}°$

b

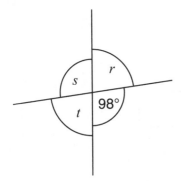

$r = \boxed{}°$ $s = \boxed{}°$ $t = \boxed{}°$

e

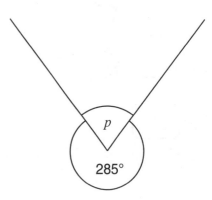

$p = \boxed{}°$

c

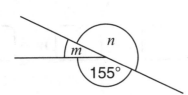

$m = \boxed{}°$ $n = \boxed{}°$

f

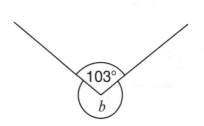

$b = \boxed{}°$

2 Calculate the missing angles in these triangles.

a

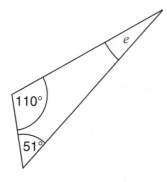

110°

51°

e

$e =$ ◻ °

d

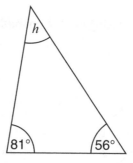

h

81°   56°

$h =$ ◻ °

b

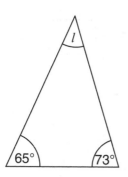

l

65°   73°

$l =$ ◻ °

e

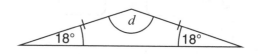

18°   *d*   18°

$d =$ ◻ °

c

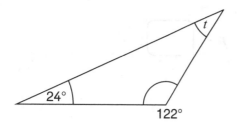

t

24°

122°

$t =$ ◻ °

f

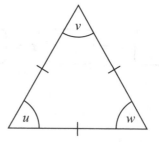

v

u   *w*

$u =$ ◻ ° $v =$ ◻ ° $w =$ ◻ °

Use the angles you are given to help you calculate the missing angles in these rectangles. Write each missing angle on the diagram.

a

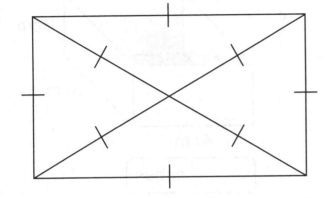

b

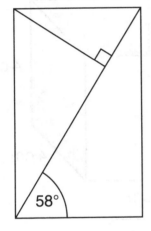

58°

c

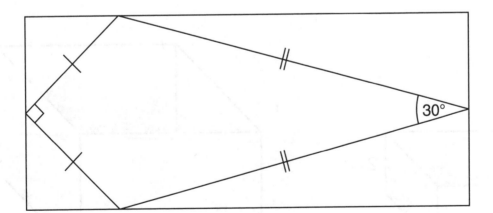

30°

4 How many 2 cm cubes will fit into each of these boxes?  2 cm

a

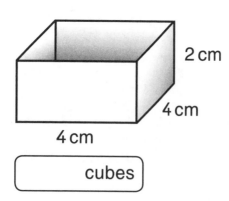

2 cm
4 cm
4 cm

cubes

b

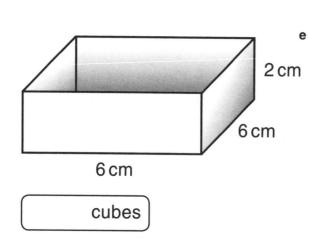

2 cm
6 cm
6 cm

cubes

c

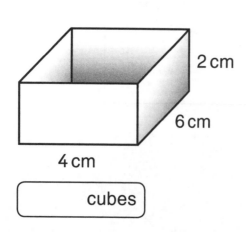

2 cm
6 cm
4 cm

cubes

d

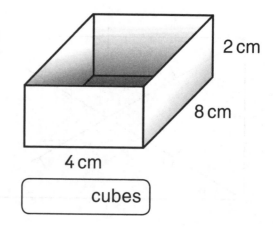

2 cm
8 cm
4 cm

cubes

e

4 cm
4 cm
4 cm

cubes

f

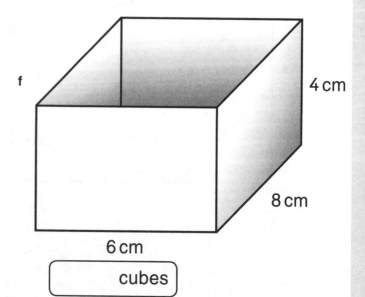

4 cm
8 cm
6 cm

cubes

5 Answer these.

a Printer paper is 30 cm × 22 cm. It comes in packs of 500 sheets. The volume of the pack is 3960 cm³. What is the height of the pack?

| cm |

c A water tank holds 192 litres. It fills at a rate of 0.2 litres per second. How long will it take to fill?

| minutes |

b A small bottle of water holds 300 ml and a large bottle holds 1.2 litres. A supermarket has the following offers:

Which is better value, large or small bottles?

| bottles |

d A cuboid has a height of 8 m. The length is half the width and the width is 3 times longer than the height. What is the volume of the cuboid?

| cm³ |

1 Look at the position of each shape on this coordinates grid.

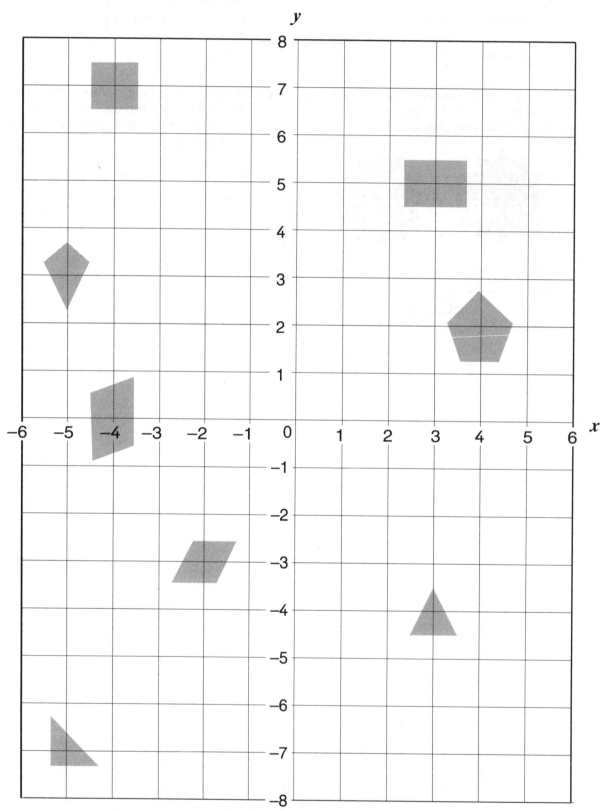

a Write the coordinates of these shapes.

(,   )

(,   )

(,   )

(,   )

b Which shapes are at the following coordinates?

(4, 2) [   ]   (-4, 0) [   ]

(–5, –7) [   ]   (–5, 3) [   ]

2 Each of the shapes on the grid above moves 1 square up and 1 square to the right.

What are the new coordinates for the shapes?

Square   (,   )   Pentagon   (,   )

Rectangle   (,   )   Right-angled triangle   (,   )

Equilateral triangle   (,   )   Parallelogram   (,   )

Rhombus   (,   )   Kite   (,   )

What do you notice?

Describe the translation of each of these.

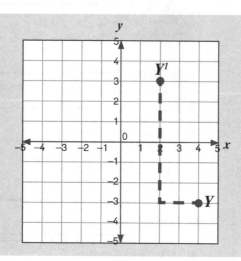

Y has been translated using -2 in a **x** direction and $+6$ in a **y** direction.

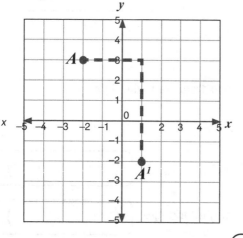

A has been translated using ☐

in a **x** direction and ☐
in a **y** direction.

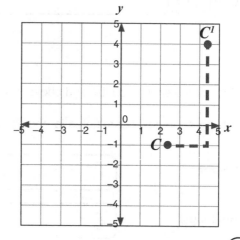

C has been translated using ☐

in a **x** direction and ☐
in a **y** direction.

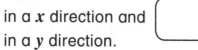

B has been translated using ☐

in a **x** direction and ☐
in a **y** direction.

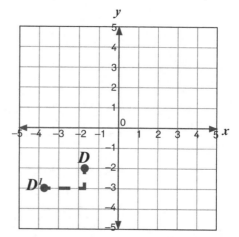

D has been translated using ☐

in a **x** direction and ☐
in a **y** direction.

4

This is a wrapping paper design.

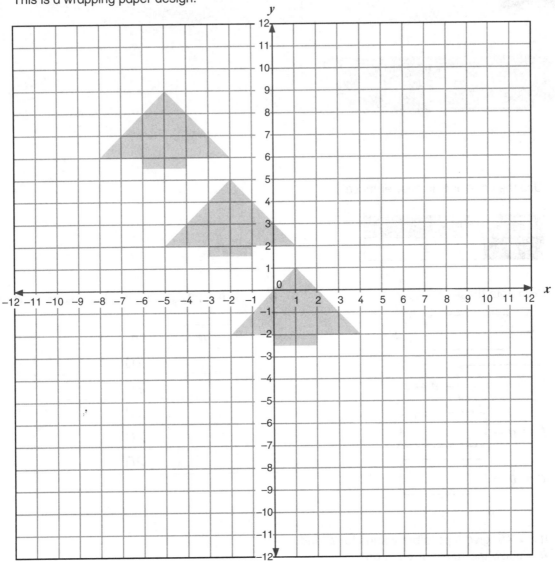

a What are the coordinates of the top tree?

c Draw the next tree in this translation pattern on the grid.

b What is the translation for the pattern?

YOU WILL NEED:
- squared paper
- ruler
- coloured pencils

a Draw a similar grid on squared paper and design your own wrapping paper.

b Record the coordinates of the vertices of your shape.

c Decide on a translation of your design and record the translation. Draw the translated pattern.

Focus on algebra

1 Use the bar model to answer these.

	a	
b	c	

a Complete these.

$a = $ ⬚ $+ c$

⬚ $- b = c$

$b = a - $ ⬚

$c + b = $ ⬚

b If $b = 20$, give 5 possible values for a and c.

a	b	c
	20	
	20	
	20	
	20	
	20	

c If $c = 45$, give 5 possible values for a and b. Follow the rules for each one.

a	b	c	Rule
		45	a is greater than 1000
		45	b is a negative number
		45	a is a square number
		45	b is a prime number
		45	$a + b$ is less than 60

2 The formula for the perimeter of a rectangle is $p = 2(l + w)$.

a What are the possible whole number values of the length and width of these rectangles?

Record the possibilities in the table.

Perimeter	Possible lengths and widths
$p = 4$ cm	$l = 1$ $w = 1$
$p = 6$ cm	
$p = 8$ cm	
$p = 10$ cm	
$p = 12$ cm	

> What do you notice about the pattern?

b Can you predict the number of possibilities for a perimeter of 14 cm?

3 Write **true** or **false** for these. Explain your answers.

a There is a whole number for n for the formula $3n + 5 = 38$

b There is a whole number for n for the formula $3n - 5 = 38$

c There is a whole number for n for the formula $\dfrac{3n}{5} = 38$

4 There are approximately 1.5 dollars to £1.

a This table shows the cost of hiring equipment for a day at a ski resort in the USA.
Complete the table.

Item	Cost in US dollars ($)	Cost in UK pounds (£)
Skis	$48	
Ski boots		£12
Helmet	$6	
Poles		£4
Snowboard	$36	

b If dollars is d and pounds is p, what is the formula for converting dollars to pounds?

c What is the cost in US dollars of hiring skis, boots, poles and a helmet for 3 days? $

1

YOU WILL NEED:
• interlocking cubes

Use interlocking cubes to make the next 3 terms for this sequence.

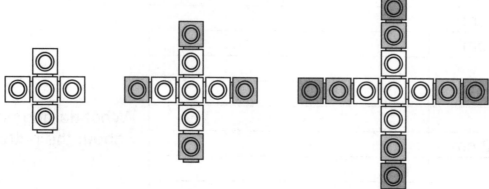

a Complete the table to show the nth term. Write the formula.

Term	Number of cubes
1	5
2	
3	
4	
5	
6	
n	

b How many cubes would be needed for

• the 20th term?

• the 50th term?

• the 100th term?

c What term would be made with 73 cubes?

d Could you make a pattern in this sequence with 130 cubes?
Explain how you know.

2 Write the next 3 numbers in these sequences.

Write a formula for the **n**th term for each sequence.

a

Term	Number
1	2
2	4
3	6
4	
5	
6	
n	

d

Term	Number
1	5
2	7
3	9
4	
5	
6	
n	

Complete the table for the next sequence.

b

Term	Number
1	3
2	5
3	7
4	
5	
6	
n	

e

Term	Number
1	
2	
3	
4	
5	
6	
n	

c

Term	Number
1	4
2	6
3	8
4	
5	
6	
n	

3 1 litre is approximately 1.75 pints. This conversion graph shows the relationship between the two measures.

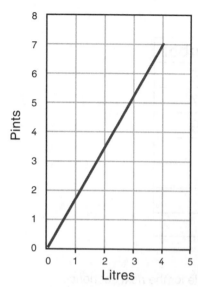

a What is the formula for this conversion of litres (*l*) to pints (*p*)? Circle the correct one.

$l = p/1.75$ $1.75l = p$ $l = p + 1.75$ $l = 1.75p$ $l/1.75 = p$

b Complete this table converting litres to pints.

Litres	2	3	4	5	6	7
Pints						

4

YOU WILL NEED:
• **ruler**

There are approximately 4.5 litres to every gallon.

a Draw a conversion graph to show this.

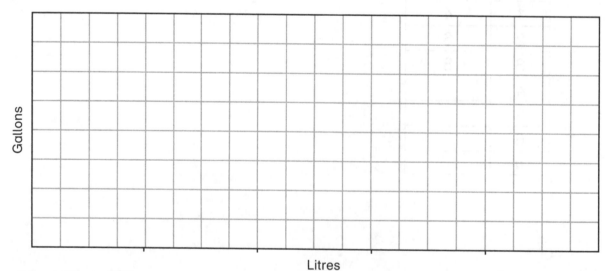

Litres

b Answer this problem.

A car travels 40 miles per gallon. Fuel costs £1.20 per litre. If £54 is spent on fuel, how many miles will the car travel on that amount?

miles

Solving more problems

1 Write operations in the boxes to make these true.

a

6 ☐ 4 ☐ 3 = 21

6 ☐ 4 ☐ 3 = 7

6 ☐ 4 ☐ 3 = 18

6 ☐ 4 ☐ 3 = 8

b

9 ☐ 3 ☐ 8 = 20

9 ☐ 3 ☐ 8 = −15

9 ☐ 3 ☐ 8 = 14

9 ☐ 3 ☐ 8 = 24

2 Calculate these. Look for mental strategies to answer each.

a $(3.6 \times 2) + (36 \times 0.2) =$ ☐

b $(4.9 \times 7) + (4.9 \times 3) =$ ☐

c $(7.1 \times 5) - (7.1 \times 3) =$ ☐

d $(4.8 \times 5) - (2.4 \times 5) =$ ☐

3 Answer these. Draw a bar model next to each to show the calculation.

a 1.37 kg of sugar is poured from a 2 kg bag into a bowl. A further 0.45 kg is poured into the bowl. How much sugar is left in the bag?

[_____] kg

c Grandma's 2 pumpkins weigh 9.36 kg together. If the heavier pumpkin is twice the weight of the lighter one, how much does each pumpkin weigh?

[_____] kg and [_____] kg

b A bag holds 4.75 kg of flour. How many bags are needed to hold 30 kg of flour?

[_____] bags

d Sam puts 2 marrows and a pumpkin into a vegetable show. The marrows each have mass 2.17 kg and the total mass of the vegetables is 6.9 kg. What is the mass of the pumpkin?

[_____] kg

4 Answer these.

a Olga drove 7.8 km from her home to go to a shop and then decided to visit a friend on the way home. The total journey was 19.5 km. How much extra distance was added to the journey by visiting her friend?

[_____] km

b A recipe to make 12 biscuits used 350 g flour, 225 g butter and 175 g sugar. However, a cook needed to make 18 biscuits. What would be the total mass of mixture for 18 biscuits?

[_____] g

c A family of 2 adults and 2 children wants to go to a theme park. Adult tickets are £37.50, child tickets are £24.75 and a family ticket for 2 adults and up to 3 children is £119.99. Which is cheaper for this family – buying separate adult and child tickets or a family ticket? How much cheaper?

[_____]

d Over a distance of 1500 metres the men's race winner took 3:27 minutes and the women's race winner took 3:50 minutes. If they started at the same time, how many metres would the women's winner be behind the men's winner when he finished?

[_____] m

1 Try different starting numbers as the denominator for each of these.

What do you notice?

$$\frac{1}{\Box} = \frac{2}{\Box} = \frac{3}{\Box} = \frac{4}{\Box} = \frac{5}{\Box} = \frac{6}{\Box}$$

$$\frac{1}{\Box} = \frac{2}{\Box} = \frac{3}{\Box} = \frac{4}{\Box} = \frac{5}{\Box} = \frac{6}{\Box}$$

$$\frac{1}{\Box} = \frac{2}{\Box} = \frac{3}{\Box} = \frac{4}{\Box} = \frac{5}{\Box} = \frac{6}{\Box}$$

2 Circle the odd one out in each set. Explain why you think it is the odd one out.

a
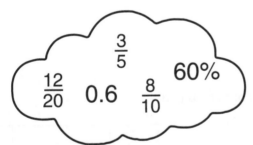
$\frac{3}{5}$ 60% $\frac{12}{20}$ 0.6 $\frac{8}{10}$

c
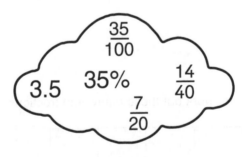
$\frac{35}{100}$ 35% $\frac{14}{40}$ 3.5 $\frac{7}{20}$

b
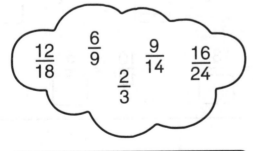
$\frac{12}{18}$ $\frac{6}{9}$ $\frac{9}{14}$ $\frac{16}{24}$ $\frac{2}{3}$

d
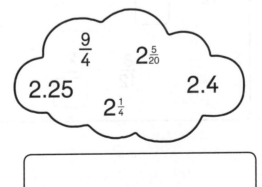
$\frac{9}{4}$ $2\frac{5}{20}$ 2.25   2.4 $2\frac{1}{4}$

3 Simplify each of these fractions. Write the number that you divide the numerator and denominator by to reduce the fraction to its simplest form.

a

$$\frac{8}{12} \begin{array}{c} \div \bigcirc \\ \\ \div \bigcirc \end{array} = \frac{\bigcirc}{\bigcirc}$$

d

$$\frac{35}{49} \begin{array}{c} \div \bigcirc \\ \\ \div \bigcirc \end{array} = \frac{\bigcirc}{\bigcirc}$$

b

$$\frac{21}{30} \begin{array}{c} \div \bigcirc \\ \\ \div \bigcirc \end{array} = \frac{\bigcirc}{\bigcirc}$$

e

$$\frac{36}{84} \begin{array}{c} \div \bigcirc \\ \\ \div \bigcirc \end{array} = \frac{\bigcirc}{\bigcirc}$$

c

$$\frac{18}{45} \begin{array}{c} \div \bigcirc \\ \\ \div \bigcirc \end{array} = \frac{\bigcirc}{\bigcirc}$$

f

$$\frac{64}{120} \begin{array}{c} \div \bigcirc \\ \\ \div \bigcirc \end{array} = \frac{\bigcirc}{\bigcirc}$$

4

a First work out these equivalent fractions. Use the bar to help you.

$$\frac{2}{12} = \frac{1}{\bigcirc} \qquad \frac{3}{12} = \frac{1}{\bigcirc} \qquad \frac{4}{12} = \frac{1}{\bigcirc} \qquad \frac{6}{12} = \frac{1}{\bigcirc}$$

$$\frac{8}{12} = \frac{2}{\bigcirc} \qquad \frac{9}{12} = \frac{3}{\bigcirc} \qquad \frac{10}{12} = \frac{5}{\bigcirc}$$

b Use the equivalents to add and subtract these fractions. Simplify the answers.

i $\quad \dfrac{1}{2} + \dfrac{1}{3} =$ ⬜

iv $\quad \dfrac{1}{2} + \dfrac{5}{6} =$ ⬜

vii $\quad \dfrac{5}{6} - \dfrac{3}{4} =$ ⬜

ii $\quad \dfrac{1}{4} + \dfrac{1}{6} =$ ⬜

v $\quad \dfrac{2}{3} - \dfrac{1}{4} =$ ⬜

viii $\quad \dfrac{2}{3} - \dfrac{1}{6} =$ ⬜

iii $\quad \dfrac{2}{3} + \dfrac{1}{4} =$ ⬜

vi $\quad \dfrac{3}{4} - \dfrac{1}{2} =$ ⬜

xi $\quad \dfrac{3}{5} - \dfrac{1}{10} =$ ⬜

 5 Answer these. Use equivalent fractions and then reduce each answer to its simplest form.

a $\quad \dfrac{2}{5} + \dfrac{3}{8} =$ ⬜ $\qquad \dfrac{2}{5} - \dfrac{3}{8} =$ ⬜

b $\quad \dfrac{3}{4} + \dfrac{3}{10} =$ ⬜ $\qquad \dfrac{3}{4} - \dfrac{3}{10} =$ ⬜

c $\quad \dfrac{4}{5} + \dfrac{7}{9} =$ ⬜ $\qquad \dfrac{4}{5} - \dfrac{7}{9} =$ ⬜

d $\quad \dfrac{2}{3} + \dfrac{1}{8} =$ ⬜ $\qquad \dfrac{2}{3} - \dfrac{1}{8} =$ ⬜

e $\quad \dfrac{9}{10} + \dfrac{1}{4} =$ ⬜ $\qquad \dfrac{9}{10} - \dfrac{1}{4} =$ ⬜

f $\quad \dfrac{7}{8} + \dfrac{1}{5} =$ ⬜ $\qquad \dfrac{7}{8} - \dfrac{1}{5} =$ ⬜

6 These are distances covered by 2 runners over 3 days.

	Monday	**Tuesday**	**Wednesday**	**Total**
Ali	$7\frac{3}{4}$ km	$6\frac{2}{3}$ km	$8\frac{2}{5}$ km	
Beth	$8\frac{1}{10}$ km	$5\frac{4}{5}$ km	$9\frac{1}{4}$ km	
Total				

a Complete the table to find the totals for each day and for each runner.

b How much further than Ali did Beth run in total?

[ ] km

c On Thursday Ali and Beth ran a total of $14\frac{1}{2}$ km. Ali ran $2\frac{1}{4}$ km further than Beth. How far did they each run on Thursday?

Ali [ ] km Beth [ ] km

1

a Find the values of c for these values of b.

	c		
b	b	b	5

b	c
1	
2	
3	
4	
5	

b Circle any equations that represent this.

$c - 5 = 3b$ $3b - 5 = c$ $3b + 5 = c$ $5 - c = 3b$ $5 = c - 3b$

2

a Draw a bar model to show the equation $m = 2n + 4$.

b The value of m is a number larger than 19 but smaller than 31. What are the possible solutions for m and n? The value of n is a **whole** number. Complete this table of results.

m						
n						

3 Complete the tables of results of possible solutions for these formulae.

a $y = 2 + x$

x	0	1	2	3	4	5	6
y							

b $y = 7 - x$

x	0	1	2	3	4	5	6
y							

c $y = 3x - 1$

x	0	1	2	3	4	5	6
y							

d $y = \dfrac{x}{2}$

x	0	1	2	3	4	5	6
y							

4 Use the results in question 3 to rewrite each formula for the value of x.

a $x = $ ☐   b $x = $ ☐   c $x = $ ☐   d $x = $ ☐

5

YOU WILL NEED:
• ruler

Plot each set of results from question 3 as a line graph. Use the values of x and y to plot the positions of each point.

a   Graph of $y = 2 + x$

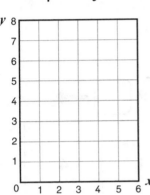

c   Graph of $y = 3x - 1$

d   Graph of $y = \dfrac{x}{2}$

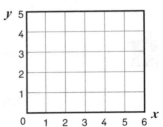

b   Graph of $y = 7 - x$

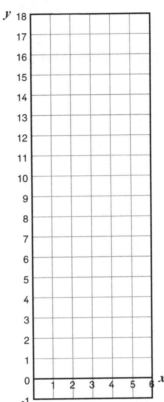

6

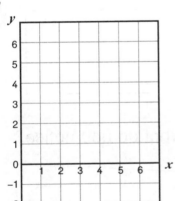

a   Write the values for x and y in this table.

x	0	1	2	3	4	5	6
y							

b   Write the formula. $y =$ ⬚

Fractions, equivalents and algebra

1 Look at the relationship between the numbers in this grid.

×4 →

1	4
3	12

×3 ↓

c

× □ →

× □ ↓

1	4
6	

Complete these.

a

×2 →

× □ ↓

1	
5	

d

×2 →

×3 ↓

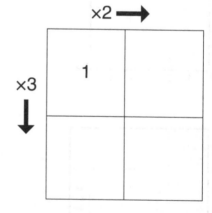

1	

What do you notice?

b

× □ →

×4 ↓

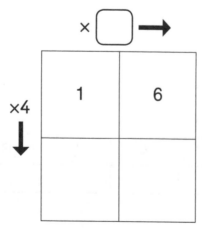

1	6

e How does this link to equivalent fractions?

Make up your own grids like those in question 1. Change the rules and the numbers. Look for patterns in your results.

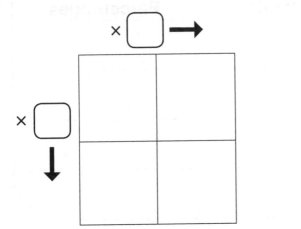

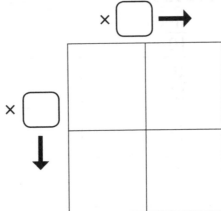

3

YOU WILL NEED:
• ruler

Explore doubling patterns with equivalent fractions to $\frac{3}{4}$.

a Shade $\frac{3}{4}$ of this rectangle.

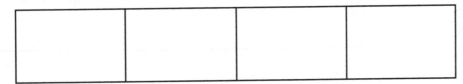

b Now draw a line with a ruler to double the number of small rectangles. Write the new fraction.

c Continue, drawing lines to double and double again the number of rectangles.

Complete this after each step.

$$\frac{3}{4} = \frac{\boxed{}}{8} = \frac{\boxed{}}{16} = \frac{\boxed{}}{32}$$

d Try this again, doubling other fractions of rectangles.

What do you notice?

4 Complete this chart to show equivalent decimals and percentages.

Fractions	Decimals	Percentages
$\frac{1}{10}$		
$\frac{3}{10}$		
$\frac{3}{5}$		
$\frac{1}{5}$		
$\frac{1}{20}$		
$\frac{1}{4}$		
$\frac{3}{4}$		
$\frac{1}{8}$		
$\frac{5}{8}$		

5 80 children were asked to choose their favourite type of fiction book. This pie chart shows the results of the survey.

Favourite types of books

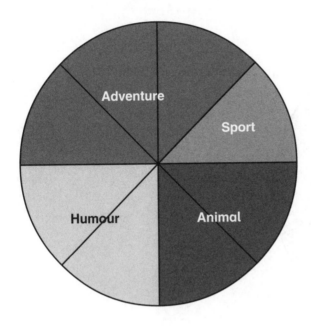

a What fraction of the children chose sports stories?

b How many children chose adventure books? ☐ children

c What percentage of the children chose humorous stories? ☐ %

b How many more children chose adventure books than sport books? ☐ children

e A further 40 children added their results to the survey. The new results showed that $\frac{1}{3}$ of the 120 children chose adventure books and 25% still chose animal stories. 10 more children chose humour books than sports books. How many children chose each type of book?

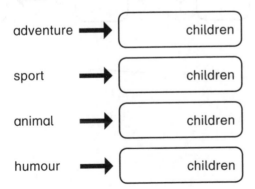

adventure ➡ ☐ children

sport ➡ ☐ children

animal ➡ ☐ children

humour ➡ ☐ children

6 This pie chart shows the results of a survey recording the first 70 vowels in a reading book. The number of vowels used were counted and recorded.

Vowels

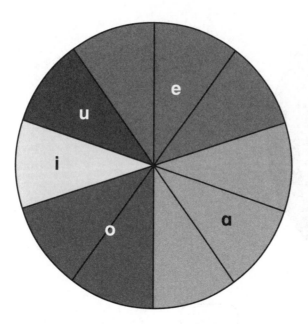

a What fraction of the vowels was the letter u? ☐

b What percentage of the vowels was the letter o? ☐ %

c What fraction of the vowels was the letter e? ☐

d There were 70 vowels altogether. How many of each vowel was counted?

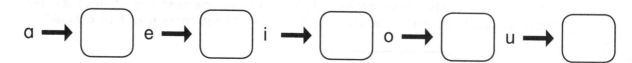

a → ☐ e → ☐ i → ☐ o → ☐ u → ☐

Try a vowel data survey with vowels in these sentences written in grey. Record the data as a pie chart and analyse the results. What fraction of the total number is each vowel? Will you do it? Try – find out!

a   Complete this tally chart. Use a protractor and a ruler to divide the circle accurately into equal parts. Record the data on the pie chart.

Vowel	Tally
a	
e	
i	
o	
u	

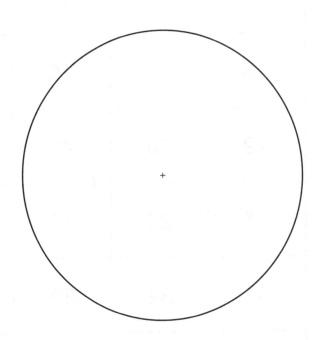

b   What fraction of the total number is each vowel?

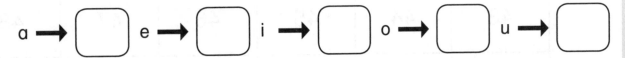

a → ☐   e → ☐   i → ☐   o → ☐   u → ☐

YOU WILL NEED:
- **counters**

Put counters on the numbers of this 60-grid for each formula.
Record the first 8 numbers and write what you notice for each sequence.

1	2	3	4	5	6
7	8	9	10	11	12
13	14	15	16	17	18
19	20	21	22	23	24
25	26	27	28	29	30
31	32	33	34	35	36
37	38	39	40	41	42
43	44	45	46	47	48
49	50	51	52	53	54
55	56	57	58	59	60

a *2n*

n	1	2	3	4	5	6	7	8
2*n*								

What do you notice?

b *2n* + 1

n	1	2	3	4	5	6	7	8
2*n* + 1								

What do you notice?

c *3n*

n	1	2	3	4	5	6	7	8
3*n*								

What do you notice?

d *3n* + 1

n	1	2	3	4	5	6	7	8
3*n* + 1								

What do you notice?

e *4n*

n	1	2	3	4	5	6	7	8
4*n*								

What do you notice?

f *4n* + 1

n	1	2	3	4	5	6	7	8
4*n* + 1								

What do you notice?

g *5n*

n	1	2	3	4	5	6	7	8
5*n*								

What do you notice?

h *5n* + 1

n	1	2	3	4	5	6	7	8
5*n* + 1								

What do you notice?

2 Calculate the area of these triangles using the formula $\frac{1}{2}(b \times h)$.

a

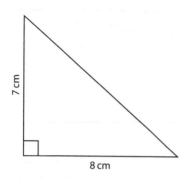

7 cm

8 cm

area = ☐ cm²

c

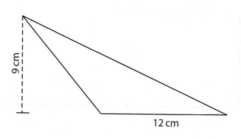

9 cm

12 cm

area = ☐ cm²

b

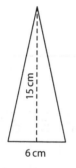

15 cm

6 cm

area = ☐ cm²

d

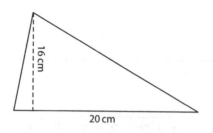

16 cm

20 cm

area = ☐ cm²

3 Calculate the area of these parallelograms using the formula $b \times h$.

a

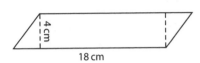

4 cm

18 cm

area = ☐ cm²

c

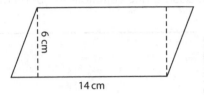

6 cm

14 cm

area = ☐ cm²

b

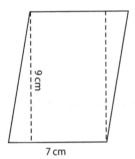

9 cm

7 cm

area = ☐ cm²

d

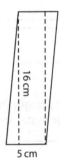

16 cm

5 cm

area = ☐ cm²

4

YOU WILL NEED:
• ruler

a   Measure and calculate the perimeter and area of this rectangle.

perimeter = [　　　　　] cm

area = [　　　　　] cm²

b   Now draw a rectangle inside this one with an area that is $\frac{1}{3}$ of the size of this rectangle. What is the perimeter and area of your rectangle?

perimeter = [　　　　　] cm

area = [　　　　　] cm²

5   Answer these.

a   The area of a square is 121 cm². What is the length of each side?

[　　　　　] cm

c   The perimeter of a square is 14 cm. What is the area?

[　　　　　] cm²

b   The perimeter of a square is 32 cm. What is the area?

[　　　　　] cm²

d   The area of a square is 6.25 cm². What is its perimeter?

[　　　　　] cm

1 Calculate the unknowns for each of these.

a $3a = 21$ $a = \boxed{}$

$3a + 6 = 21$ $a = \boxed{}$

c $4c - 5 = 39$ $c = \boxed{}$

$4c + 3 = 39$ $c = \boxed{}$

b $12 = 8 + b$ $b = \boxed{}$

$12 = 8 + 2b$ $b = \boxed{}$

2 Circle the value that solves each equation. Show how you worked each out.

a $4y + 3 = 21 + y$

3   4   6   8   9

Working:

c $\dfrac{p}{3} + 2 = \dfrac{2p}{4}$

3   6   9   12   15

Working:

b $2m - 5 = 3m - 14$

15   9   8   7   4

Working:

d $3n - 8 = n + 6$

11   10   9   8   7

Working:

3 Answer these.

a Is it **true** or **false** that there is a whole number that satisfies the equation

$$20 - n = 3n$$

true ☐   false ☐

Show your reasoning.

☐

b Is it **true** or **false** that there is a whole number that satisfies the equation

$$2n - 1 = \frac{n}{2}$$

true ☐   false ☐

Show your reasoning.

☐

4 Solve these problems.

a A fruit smoothie drink has 2 bananas for every 5 oranges. The drink was made with 35 pieces of fruit. How many bananas and how many oranges were used for this smoothie?

☐ bananas and ☐ oranges

b Gita had £1.40 and Luke had twice this amount. Together they wanted to buy a game that cost £3.99. Do they have enough money to buy the game?

☐

c A lorry travels 57 km from Luton to London, and then back again every day. The lorry has enough fuel in the tank for 300 km. How many return journeys from Luton to London and back again can the lorry complete before needing to fill up with fuel?

☐

d A biscuit recipe uses 250 g butter, half this amount of sugar and twice the amount of flour than butter. This recipe makes 25 cookies. What is the mass of each cookie?

☐ g

e A coat was reduced by 20% in a sale. The sale price was £42.00. The next week the coat was back to the full price but had a sign saying £15 off. What is the latest price of this coat?

£ ☐

f Jake had 5 bags of balloons. Each bag had 12 balloons. Jake kept a quarter of the balloons for himself and then shared the rest equally between 5 friends. How many balloons did Jake have and how many did each friend get?

Jake ☐   Each friend ☐

5 Answer these as equations. Harry collects dinosaur stickers. He has a box of n stickers.

a He buys 6 more stickers. How many has he now got in his box?

c Harry counts all his stickers and he has 48 in his box. Write this as an equation from the information given.

b A friend then gives Harry all her spare stickers. This doubles the number of stickers Harry has in his box. How many stickers has he now got?

d How many stickers did Harry have to begin with?

$n =$

13a Using long division

1 Answer these. Use the first fact to help answer the others.

a $3 \times \boxed{} = 21$

$21 \div 3 = \boxed{}$

$210 \div 3 = \boxed{}$

$21 \div 30 = \boxed{}$

$2.1 \div 3 = \boxed{}$

$210 \div 300 = \boxed{}$

b $9 \times 4 = \boxed{}$

$\boxed{} \div 4 = 9$

$\boxed{} \div 40 = 9$

$\boxed{} \div 4 = 900$

$\boxed{} \div 4 = 0.9$

$\boxed{} \div 40 = 90$

c $\boxed{} \times 8 = 48$

$48 \div 8 = \boxed{}$

$480 \div 8 = \boxed{}$

$48 \div 80 = \boxed{}$

$4.8 \div 8 = \boxed{}$

$480 \div 80 = \boxed{}$

2 Answer these. Write the remainders as numbers, fractions and decimals.

$38 \div 4 = \boxed{9 \ r2}$

$38 \div 4 = \boxed{9\tfrac{1}{2}}$

$38 \div 4 = \boxed{9.5}$

a $57 \div 5 = \boxed{}$

$57 \div 5 = \boxed{}$

$57 \div 5 = \boxed{}$

c $129 \div 6 = \boxed{}$

$129 \div 6 = \boxed{}$

$129 \div 6 = \boxed{}$

b $74 \div 8 = \boxed{}$

$74 \div 8 = \boxed{}$

$74 \div 8 = \boxed{}$

d $153 \div 12 = \boxed{}$

$153 \div 12 = \boxed{}$

$153 \div 12 = \boxed{}$

3 Answer these. Write remainders as **fractions**.

a   15 ⎾4825

c   24 ⎾4712

b   18 ⎾6303

d   32 ⎾7196

4   Answer these.

a   Water pipes come in 3 m lengths. A new supply of water is needed for a new house which is 3809 m away from the main water supply. How many lengths of 3 m pipe will be needed to take the water supply to this house?

     [lengths]

b   A factory makes 7275 pencils a day. They are put in boxes of 8 to sell to schools. How many full boxes of pencils are made each day?

     [boxes]

c   A tractor brings 1462 kg of potatoes from the fields to the barn. The potatoes are put into 25 kg sacks ready for sale. How many 25 kg sacks of potatoes can be filled from this load of potatoes?

     [sacks]

d   Sara has worked out that today her Grandad is 1143 months old. How old is her Grandad in years?

     [years]

e   Lee is in a sponsored race and he wants to run more than 2 km. The whole school is running lengths of the football pitch. The pitch is 90 m long. How many full lengths must Lee run to know he has run more than 2 km?

     [lengths]

f   There are 1257 children in a large school. Two-thirds of them travel home by bus. A bus has 53 seats. How many buses are needed to take the children home?

     [buses]

Write in the missing digits.

a

$$
\begin{array}{r}
5\ \boxed{} \\
1\ \boxed{}\ \overline{)\ 7\ \ 9\ \ 5}
\end{array}
$$

e

$$
\begin{array}{r}
1\ \ 8 \\
2\ 6\ \overline{)\ \boxed{}\ 6\ \boxed{}}
\end{array}
$$

b

$$
\begin{array}{r}
6\ \ 7 \\
1\ 4\ \overline{)\ 9\ \boxed{}\ \boxed{}}
\end{array}
$$

f

$$
\begin{array}{r}
4\ \ 6 \\
1\ \boxed{}\ \overline{)\ 8\ \boxed{}\ 4}
\end{array}
$$

c

$$
\begin{array}{r}
1\ \ 2 \\
\boxed{}\ \boxed{}\ \overline{)\ 5\ \ 7\ \ 6}
\end{array}
$$

g

$$
\begin{array}{r}
2\ \ 7 \\
2\ 6\ \overline{)\ 7\ \boxed{}\ \boxed{}}
\end{array}
$$

d

$$
\begin{array}{r}
2\ \boxed{} \\
\boxed{}\ 1\ \overline{)\ 9\ \ 4\ \ 3}
\end{array}
$$

h

$$
\begin{array}{r}
\boxed{}\ 2 \\
2\ 1\ \overline{)\ 8\ \ 8\ \boxed{}}
\end{array}
$$

 6

YOU WILL NEED:
- digit cards 1–9

All the digits in this division are different.

$$
4\ 6\ 8 \div 1\ 2 = 3\ 9
$$

Use digit cards 1 to 9 to find other divisions that use 7 different digits.

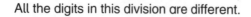

$$
\boxed{}\boxed{}\boxed{} \div \boxed{}\boxed{} = \boxed{}\boxed{}
$$

1 Write the missing operations to make these true.

$$+ \quad - \quad \times \quad \div$$

a 8 ☐ 4 ☐ 2 = 10

d 8 ☐ 4 ☐ 2 = 34

b 8 ☐ 4 ☐ 2 = 30

e 8 ☐ (4 ☐ 2) = 48

c 8 ☐ 4 ☐ 2 = 64

f 8 ☐ (4 ☐ 2) = 4

2 Freeway Coach Company offers these discounts for group bookings of passengers.

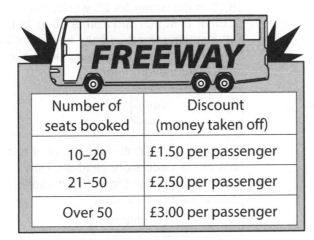

Number of seats booked	Discount (money taken off)
10–20	£1.50 per passenger
21–50	£2.50 per passenger
Over 50	£3.00 per passenger

Calculate the total cost of each of these bookings. Show how you worked them out.

a 24 seats @ £18 each

Total cost is £ ☐

c 32 seats @ £16.50 each

Total cost is £ ☐

b 15 seats @ £39 each

Total cost is £ ☐

d 79 seats @ £26.50 each

Total cost is £ ☐

3 Answer these.

a Paving slabs are 0.8 m in length. A path is made using 14 slabs in a row. How long is the path?

☐ m

b If these paving slabs are square, what is the area of the path?

☐ m²

c It is 196 km from Leeds to London. A bus travels there and back every day for 1 week. How far does the bus travel in 1 week?

☐ km

d The bus travels at an average speed of 49 km per hour. How long does it take to travel from Leeds to London?

☐ hours

e There are 84 guests at a wedding. The chef makes 260 sandwiches and expects the guests to eat 3 sandwiches each. How many sandwiches will be left over?

☐ sandwiches

f The guests are seated in equal numbers at 7 tables. How many sandwiches will be placed on each table?

☐ sandwiches on each table

g A zoo is open every day from 10:00 to 16:00 in the winter for 25 weeks from October to March and from 9:00 to 18:00 in the summer for 26 weeks from April to September. How many hours is the zoo open in total every year?

☐ hours

h The zoo has 142 000 visitors a year and 60% of the visitors are children. How many children visit the zoo in a year?

☐ children

4 Entrance to a castle is £6 per adult and £4 per child. A group of 8 people pay £38 in total. How many adults and children are there in the group?

☐ children   ☐ adults

Complete the table to help. Explain your reasoning.

Number	£6 adult	£4 children
1		
2		
3		
4		
5		
6		
7		
8		

_____

_____

_____

_____

1 Answer these.

$$3 \times \frac{1}{2} = \boxed{\frac{1}{2}} + \boxed{\frac{1}{2}} + \boxed{\frac{1}{2}} = \boxed{\frac{3}{2}} = \boxed{1\frac{1}{2}}$$

a $4 \times \frac{1}{2} = \boxed{} + \boxed{} + \boxed{} + \boxed{} = \boxed{} = \boxed{}$

b $5 \times \frac{1}{2} = \boxed{} + \boxed{} + \boxed{} + \boxed{} + \boxed{} = \boxed{} = \boxed{}$

c $6 \times \frac{1}{2} = \boxed{} + \boxed{} + \boxed{} + \boxed{} + \boxed{} + \boxed{} = \boxed{} = \boxed{}$

d $4 \times \frac{1}{4} = \boxed{} + \boxed{} + \boxed{} + \boxed{} = \boxed{} = \boxed{}$

e $4 \times \frac{3}{4} = \boxed{} + \boxed{} + \boxed{} + \boxed{} = \boxed{} = \boxed{}$

f $4 \times \frac{1}{5} = \boxed{} + \boxed{} + \boxed{} + \boxed{} = \boxed{}$

g $4 \times \frac{3}{5} = \boxed{} + \boxed{} + \boxed{} + \boxed{} = \boxed{} = \boxed{}$

What do you notice?

2 Calculate the answers to these. Simplify if possible.

$$\frac{1}{3} \times \frac{1}{4} = \boxed{\frac{1}{12}}$$

	$\frac{1}{4}$	$\frac{1}{4}$	$\frac{1}{4}$	$\frac{1}{4}$
$\frac{1}{3}$				
$\frac{1}{3}$				
$\frac{1}{3}$				

a $\quad \frac{1}{6} \times \frac{1}{4} = \boxed{}$

d $\quad \frac{2}{3} \times \frac{1}{2} = \boxed{}$

g $\quad \frac{1}{3} \times \frac{3}{4} = \boxed{}$

b $\quad \frac{1}{6} \times \frac{1}{2} = \boxed{}$

e $\quad \frac{2}{3} \times \frac{1}{4} = \boxed{}$

h $\quad \frac{1}{6} \times \frac{3}{4} = \boxed{}$

c $\quad \frac{1}{3} \times \frac{1}{2} = \boxed{}$

f $\quad \frac{2}{3} \times \frac{3}{4} = \boxed{}$

i $\quad \frac{5}{6} \times \frac{3}{4} = \boxed{}$

 3 This shows that $\quad \frac{1}{3} \div 2 = \frac{1}{6}$

$\frac{1}{3}$	$\frac{1}{3}$	$\frac{1}{3}$
$\frac{1}{6}$		

Use the bar models to help answer these. Simplify if possible.

a $\quad \frac{1}{3} \div 3 =$

$\frac{1}{3}$	$\frac{1}{3}$	$\frac{1}{3}$

b $\quad \frac{1}{3} \div 4 =$

$\frac{1}{3}$	$\frac{1}{3}$	$\frac{1}{3}$

c $\frac{2}{3} \div 4 =$

$\frac{1}{3}$	$\frac{1}{3}$	$\frac{1}{3}$

d $\frac{1}{4} \div 3 =$

$\frac{1}{4}$	$\frac{1}{4}$	$\frac{1}{4}$	$\frac{1}{4}$

4 Martha was given £60 for her birthday. This is how she spent her money in order.

a Write how much she has left each time.

She spent ...	Money remaining
$\frac{1}{2}$ of her money on a coat	
$\frac{1}{3}$ of what was left on a tennis racket	
$\frac{1}{4}$ of what was left on a pen	
$\frac{1}{5}$ of what was left on a book	
$\frac{1}{6}$ of what was left on a hat	
$\frac{2}{5}$ of what was left on a magazine	
$\frac{1}{4}$ of what was left on a box of chocolates	
$\frac{2}{5}$ of what was left on a pencil	

b How much did she have left at the end?   £

Nets, angles and coordinates

1   Convert these units of measure.

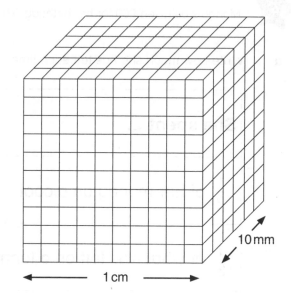

10 mm

1 cm

a $1\,cm =$ [mm]

$1\,cm^2 =$ [mm^2]

$1\,cm^3 =$ [mm^3]

b   [cm] $= 40\,mm$

[cm^2] $= 400\,mm^2$

[cm^3] $= 4000\,mm^3$

c   [cm] $= 50\,mm$

[cm^2] $= 500\,mm^2$

[cm^3] $= 5000\,mm^3$

e   [m] $= 200\,cm$

[m^2] $= 200\,cm^2$

[m^3] $= 200\,cm^3$

d $1\,m =$ [cm]

$1\,m^2 =$ [cm^2]

$1\,m^3 =$ [cm^3]

f $6\,m =$ [cm]

$60\,m^2 =$ [cm^2]

$600\,m^3 =$ [cm^3]

YOU WILL NEED:
- cubes (multilink or similar)

Use 1 cm cubes to make each of these cubes.

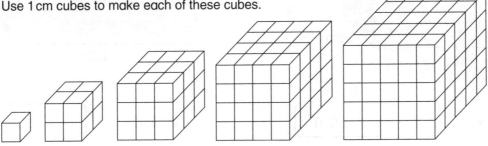

a   Record their volume and surface areas in this table.

Side length (cm)	Volume (cm³)	Surface area (cm²)
1		
2		
3		
4		
5		

b   What do you notice?

c   Can you predict the surface area of a cube with sides of 10 cm

| cm² |

3   Calculate the volume and surface area of each cuboid.

a

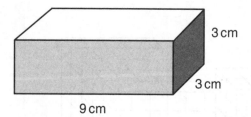

3 cm

3 cm

9 cm

volume = [   ] cm³

surface area = [   ] cm²

d

0.4 m

0.8 m

2.2 m

volume = [   ] m³

surface area = [   ] m²

b

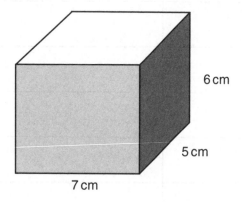

6 cm

5 cm

7 cm

volume = [   ] cm³

surface area = [   ] cm²

e

1.5 m

1.5 m

1.5 m

volume = [   ] m³

surface area = [   ] m²

c

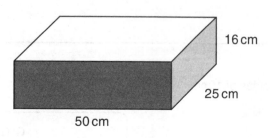

16 cm

25 cm

50 cm

volume = [   ] cm³

surface area = [   ] cm²

f

100 mm

90 mm

85 mm

volume = [   ] mm³

surface area = [   ] mm²

4 How many 125 cm³ cubes will fit into each of these boxes?  125 cm³

a

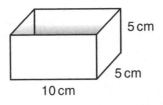
5 cm
5 cm
10 cm

[cubes]

b

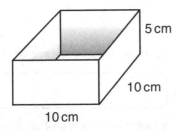
5 cm
10 cm
10 cm

[cubes]

c

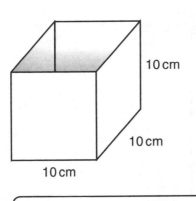
10 cm
10 cm
10 cm

[cubes]

d

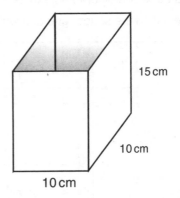
15 cm
10 cm
10 cm

[cubes]

e

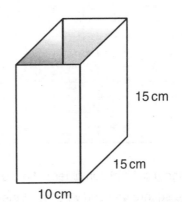
15 cm
15 cm
10 cm

[cubes]

f

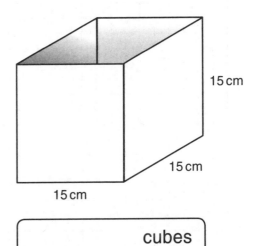
15 cm
15 cm
15 cm

[cubes]

g What is the volume of the **largest** box?

[cm³]

5 A box for a teapot is a cube with sides of 20 cm.

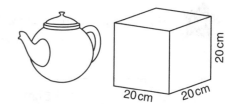

a Design a **cube** container to hold 64 teapot boxes.
Draw the net on the square grid and write in all the dimensions.

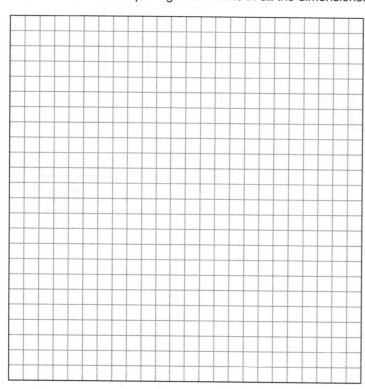

What is the surface area of your cube?

What is the volume of your cube?

b Now design a **cuboid** container to hold 64 teapot boxes. Draw the net on the square grid and write in all the dimensions.

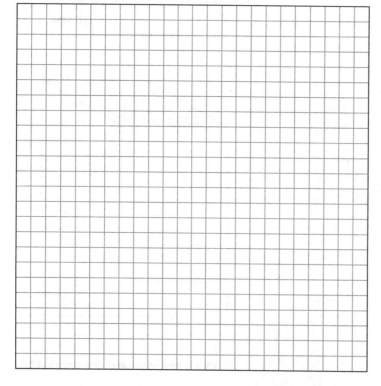

What is the surface area of your cuboid?

What is the volume of your cuboid?

c Which container – the cube or the cuboid – has the smallest surface area?

1 Calculate these missing angles.

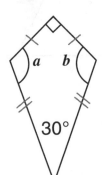

30°

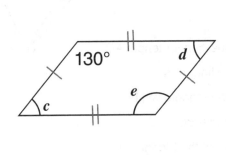

130°

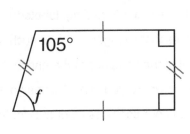

105°

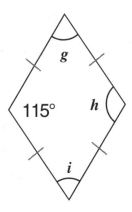

115°

a

°

b

°

c

°

d

°

e

°

f

°

g

°

h

°

i

°

2

Follow these instructions to construct different triangles, ABC, using compasses and a ruler.

- Draw a base line exactly 5 cm long, labeled AB.
- Open compasses to the lengths given in each question for length AC.
- Place the point of the compass at the end of the line, at A.
- Draw a small arc where you think the apex of the triangle will be.
- Repeat this with your compass point at B for length BC.
- The intersection of the two arcs is the third point of the triangle, at C.

a AC = 5 cm, BC = 5 cm

c AC = 3 cm, BC = 4 cm

b AC = 8 cm, BC = 8 cm

d AC = 4 cm, BC = 6 cm

e What do you notice about any of the triangles?

Work out the missing angles in these shape patterns.

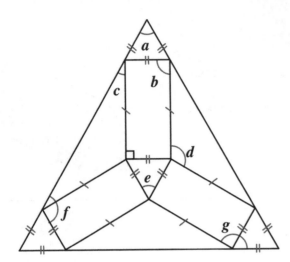

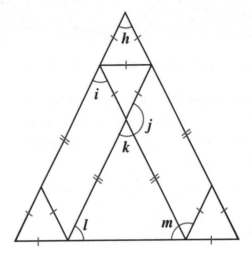

a   [⚬]

b   [⚬]

c   [⚬]

d   [⚬]

e   [⚬]

f   [⚬]

g   [⚬]

h   [⚬]

i   [⚬]

j   [⚬]

k   [⚬]

l   [⚬]

m   [⚬]

4

YOU WILL NEED:
- compasses
- ruler
- card
- scissors
- glue or sticky tape

Design your own chocolate box in the shape of a triangular prism.

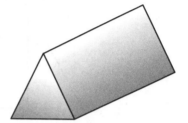

- Use compasses and a ruler to construct a net of your package.
- Write the angles and lengths of each side of your shape.
- Make the net on card, and include tabs for gluing.
- Make your box.

Sketch your net here.

What is the surface area of your triangular prism?

1 Complete the tables of results for these function machines.

x and y are the two unknowns for IN and OUT.

a $\triangle = \square + 3$
    $y = x + 3$

IN x	0	1	2	3	4	5
OUT y						

b $\triangle = 2\square + 3$
    $y = 2x + 3$

IN x	0	1	2	3	4	5
OUT y						

c $\triangle = 2\square - 3$
    $y = 2x - 3$

IN x	0	1	2	3	4	5
OUT y						

d $\triangle = 3\square + 3$
    $y = 3x + 3$

IN x	0	1	2	3	4	5
OUT y						

2 Write the equations for these tables of results.

a

IN x	0	1	2	3	4	5
OUT y	3	4	5	6	7	8

$y =$ _____

b

IN x	0	1	2	3	4	5
OUT y	−5	−4	−3	−2	−1	0

$y =$ _____

c

IN x	0	1	2	3	4	5
OUT y	2	4	6	8	10	12

$y =$ _____

d

IN x	0	1	2	3	4	5
OUT y	−1	2	5	8	11	14

$y =$ _____

3 This shows the graph for the equation $y = x + 3$

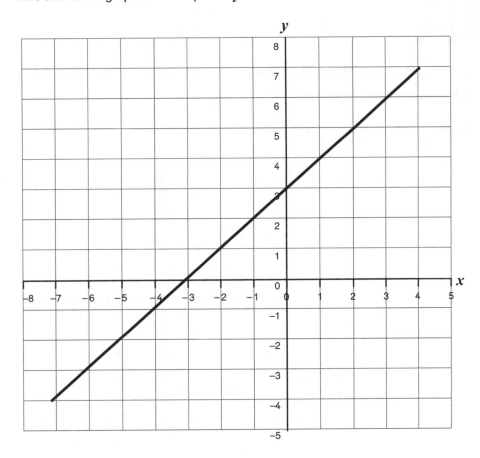

a   If $x = 3$, then $y =$ [ ]

b   If $x = 0$, then $y =$ [ ]

c   If $x = -4$, then $y =$ [ ]

d   If $x = 2$, then $y =$ [ ]

e   If $y = 0$, then $x =$ [ ]

f   If $y = -7$, then $x =$ [ ]

YOU WILL NEED:
• ruler

Look at your equations and tables of results for question 2. Draw graphs for each equation, joining the coordinates to make a straight line.

a $y = \boxed{}$

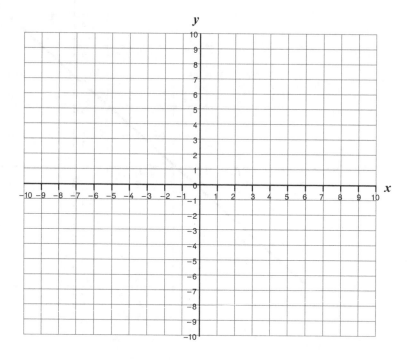

b $y = \boxed{}$

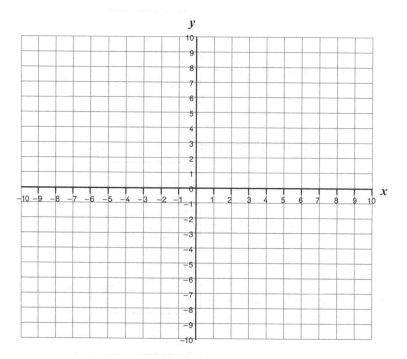

c $y = \boxed{}$

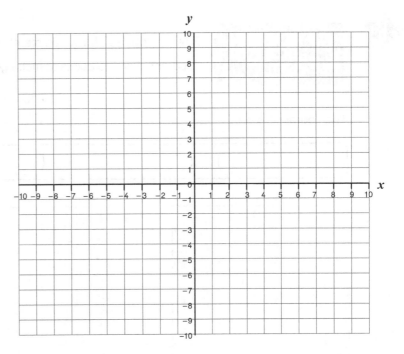

d $y = \boxed{}$

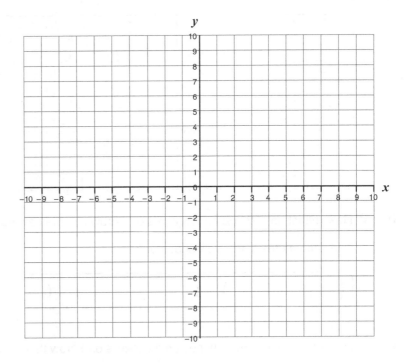

5 Compare your graphs above. Write 3 things you notice about them.

6

YOU WILL NEED:
• ruler

Plot the vertices of a quadrilateral of your choice on this grid in the first quadrant. Join the points to make your shape.

a The coordinates are:

(,   ) (,   ) (,   ) (,   )

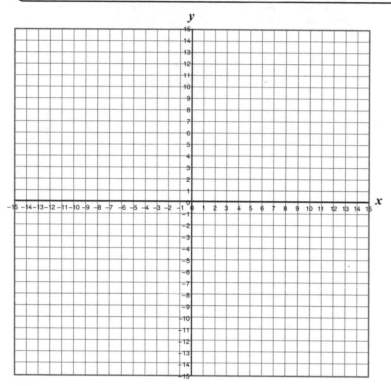

b Now reflect your shape on the **y**-axis and draw it in the new position.

The coordinates of this new shape are:

(,   ) (,   ) (,   ) (,   )

c Now reflect both these shapes on the **x**-axis and draw them in their new positions.

The coordinates of these shapes are:

(,   ) (,   ) (,   ) (,   )

(,   ) (,   ) (,   ) (,   )

d What do you notice about the coordinates of the four shapes?

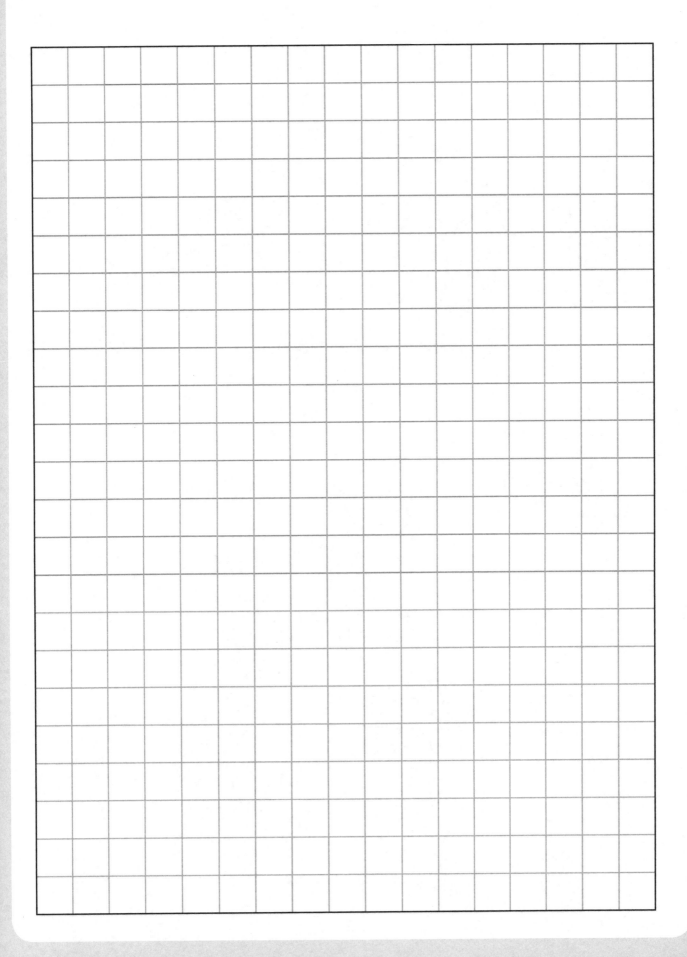

My jottings

Rising Stars Mathematics includes high-quality
Textbooks, Teacher's Guides, Practice Books, online tools
and CPD to provide a comprehensive mastery
programme for mathematics.

For more information on the complete range, visit
www.risingstars-uk.com/rsmathematics.

RISING STARS
Mathematics

Join Ali and Eva as you work through this book to practise all the new skills you have learnt in class. There are questions for every unit in Year 6, to help you master each concept.

ISBN 978-1-78339-819-5

9 781783 398195 >

RISING★STARS

For more information please call 01235 400 555
www.risingstars-uk.com

Follows the NCETM textbook guidance